Thomas Fanning Wood, Gerald McCarthy

Wilmington Flora

A list of Plants Growing about Wilmington, North Carolina, with Date of

Flowering.

Thomas Fanning Wood, Gerald McCarthy

Wilmington Flora
A list of Plants Growing about Wilmington, North Carolina, with Date of Flowering.

ISBN/EAN: 9783337270032

Printed in Europe, USA, Canada, Australia, Japan

Cover: Foto ©berggeist007 / pixelio.de

More available books at **www.hansebooks.com**

WILMINGTON FLORA

A LIST OF

PLANTS GROWING ABOUT WILMINGTON,
NORTH CAROLINA, WITH DATE
OF FLOWERING.

WITH

A MAP OF NEW HANOVER COUNTY.

BY

THOMAS F. WOOD and GERALD McCARTHY.

RALEIGH :
EDWARDS, BROUGHTON & CO., POWER PRINTERS AND BINDERS.
1886.

WILMINGTON FLORA.

THOMAS F. WOOD AND GERALD McCARTHY.

INTRODUCTION.

The growing interest in botany during the last ten years has created the necessity for a manual for local work in botany, especially as regards the Cape Fear region.

The pamphlet, published by Dr. Curtis in 1832, has long since become a rarity, and was not up to the recent date as regards the newer discoveries, and abounded in old synonyms. An attempt was first made to rewrite this catalogue, substituting the accepted synonyms of to-day, but it was found best to re-write it entire. This pamphlet, therefore, has for its basis the work of Dr. Curtis. To it has been added all the plants which he subsequently added in his catalogue of the plants of the State, and the few other plants which have been added since that work was issued.

We have given the date of flowering of plants, founded as much on personal observation as possible, but still we feel that this part of the.work can be much improved. For the sake of many beginners, we have also added the color of flowers, knowing, of course, that to the scientific botanist, such a slight aid would be so imperfect as to be of little use to him; but, for beginners, who are puzzling over plant analyses for the first time, such a slight knowledge as the color of a flower would often be of help. Furthermore, amateur botanists, of whom there is an increasing number all over the State, appreciate such slight aids, and for these considerations we think the trouble has been well spent.

We have added a map of the Cape Fear region which is accurately drawn to a scale (one mile to ½ inch), and will be found valuable to botanists and others. The basis of this map was one drawn by Capt. W. H. James, of the Confederate States Engineers, after a complete topographical survey during the war. Corrections and ad. ditions have been made to date, as nearly as possible, excepting

that the names of the proprietors of farms have been in many instances unaltered. It is easy enough, with this map, to go to any
locality in the county, and botanists and amateurs, seeking the quite
numerous local plants of this favored region, could hardly go amiss.

Dr. Curtis' original catalogue enumerated species found within a
radius of two miles of Wilmington; but we have extended our search
over the entire county, even extending to what is now a part of
Pender county, especially including the region of Rocky Point, and
of Smithville and Smith's Island, and the part of Brunswick county immediately adjacent to Wilmington.

" Wilmington is situated in latitude 34° 17', and longitude 78° 10',
about thirty miles from the mouth of the Cape Fear river, on which
it stands, and eight miles from the sea, in an easterly direction. Its
precise elevation above the ocean I have not learned,* but it is so
small as to deserve little or no consideration in regard to botanical
geography. Indeed much of the lowland in the vicinity is but little
above the level of the ocean. The climate may be pretty well de-
termined from the following table of temperature, made from observations taken in 1832. The thermometer was placed in the shade
on the north side of a house. Observations were taken six times a
day from 8 and 9 o'clock A. M. to 11 P. M. Fractions omitted:

	Jan.	Feb.	Mar	Apl.	May	Ju'e	July	Aug	Sept	Oct.	Nov	Dec.
Maximum	72	75	80	87	....	97	86	83	85	82	77	68
Minimum	18	32	25	42	.....	60	68	66	63	40	36	27
Medium..........	46	55	56	65	....	77	72	79	75	66	57	50
Rainy days	4	8	4	8	...	5	9	14	12	4	6	5

†Temperature for Year 1885.

	Jan.	Feb.	Mar	Apl.	May	Ju'e	July	Aug	Sept	Oct.	Nov	Dec.
Maximum	74	70	71	84	88	93	94	94	89	81	80	71
Minimum	23	20	26	37	51	58	62	60	52	42	30	27
Mean	49	45	49	62	70	76	81	80	73	62	55	50
Rainy days	17	9	11	6	14	7	15	15	7	10	10	9

*The elevation of Wilmington is twenty-five feet above sea level.

†This report was obtained from the U. S. Signal Service officer at Wilmington. The number of rainy days includes all days on which as much as .01 of
an inch fell, which probably accounts for the excess of rainy days in 1885.

I have not materials for forming an accurate Floral Calendar for Wilmington, but the following notices show the flowering time of a few plants in the spring of 1832:

Daffodils in flower,	Feb. 1	Red Cedar and Elm	Feb. 1
Red Maple	" 8	Jonquils	" 10
Peach and Plum	" 12	Cercis Canadensis	" 26
Flowering Almond	" 26	Epigæa repens	" 26
Phlox subulata	" 27	Luzula Campestris	" 27
Vaccinium corymbosum	March 1	Viola cucullata and lanceolata	
Cardamine Virginica	" 6		March 6
Thlaspi bursa-pastoris	" 10	Bayberry (Myrica)	" 10

In 1831, Daffodils blossomed January 1st, and the White Hyacinth at Christmas, which is about their usual period of flowering." —*Moses A. Curtis, A. M., in " Catalogue of Plants Growing Spontaneously Around Wilmington."*

The following list of plants, with their dates of flowering in 1885, is taken from the note book of Miss Hettie E. Watson, Secretary of " The Curtis Club," of Wilmington:

Alder,	February 1	Cydonia Japonica,	February 20
Chickweed	" 2	Pyxidanthera barbulata,	" 26
Blue Violet,	" 2	Short Leaf Pine stam,	" 26
White Violet,	" 3	Short Leaf Pine pistillate	" 26
Arbutus,	" 3	Oxalis yellow,	March 7
Jessamine, yellow	" 3	Water cress,	" 7
Huckleberry (*Vaccinium*		Pepper grass (*Lepidium*	
corymbosum.)	" 3	*Virginicum*.	" 23
Chrysogonum Virginianum	" 3	Nasturtium,	" 23
Bartonia verna,	" 3	Willow catkin, sterile.	" 26
Cedar,	" 5	Willow fertile,	" 26
Long Leaf Pine stam,	" 5	Strawberry, yellow, (*Fraga-*	
Long Leaf Pine pistillate,	" 16	*ria Indica*,)	" 26
Schallott,	" 16	Strawberry, white, (*F. Vir-*	
Gill,	" 16	*giniana*),	" 26
Red bud,	" 16	Peach,	" 26
Cypress,	" 16	Sassafras,	" 27
Red Maple,	" 16	Golden Club, (*Orontium*),	" 28
Red Cedar,	" 16	Rubus trivialis,	April 1
Elm,	" 16	Euphorbia,	" 1
Prunus Carolinianum,	" 19	Willow Oak, (*Quercus Phel-*	
Mistletoe,	" 19	*los*,	" 4

Scrubby Oak, (*Q. Catesbæi*)..April 4
Live Oak, (*Q. Virens*) ...... " 4
Water Oak, (*Q. Aquatica*)... " 4
Sorrel wood, (*Oxydendrum ar-*
    boreum) .... " 4
Collard, ... " 4
Blood root, (*Sanguinaria Can-*
    adensis) .... " 4
Ranunculus acris, ... " 5
Rubus villosus, ... " 5
Blue Plum, ... " 5
Azalea nudiflora, ... " 5
Woodbine, (*Lonicera semper-*
    virens) .... " 6
Pear, ... " 6
Low Laurel, (*Kalmia angus-*
    tifolia) .... " 6
Cratægus, ... " 7
Euphorbia commutata, ... " 7
White Clover, ... " 7
Ilex glabra, ... " 7
Linaria Canadensis, ... " 7
Pontederia, ... " 9
Sweet Gum, ... " 9
Horse Radish, ... " 9
Mulberry, ... " 9
Myrtle, pistillate, ... " 9
Myrtle, stam, ... " 9
Iris verna, ... " 9
Elder, small, ... " 9
Turnip, ... " 10
Buckeye, ... " 10
Dogwood, ... " 10
Wild Olive, ... " 10
Quince, ... " 10
Crab Apple, ... " 10
Horse Apple, ... " 10
Apricot, ... " 10
Running Huckleberry, (*Vac-*
    cinium crassifolium).... " 10

Clematis crispa, ...April 11
Cyrilla racemiflora, ... " 11
Pinguicula elatior, ... " 1
Salix Nigra, ... " 19
Paulownia, ... " 20
Asarum, ... " 20
Sysyrinchium Bermudiana,. " 20
Uvularia, ... " 20
Crowfoot, ... " 21
Campanula, ... " 21
Wistaria, ... " 21
Wild Cherry, (*Prunus sero-*
    tina, ... " 21 .
Lespedeza, ... " 21
Leptocaulis divaricatus, ... " 22
Lupine, purple, ... " 22
Pea, (English) .... " 25
Irish Potatoe, ... " 25
Butter wort, (*Pinguicula pu-*
    milla,) .... " 25
Rosa Carolina, ... " 27
Sarracenia flava, ... " 27
Drosera brevifolia, ... " 29
Wild phlox, ... " 29
Melia azederach, ... " 29
Ilex opaca, ... " 29
Walnut, ... " 29
Chinquapin, ... " 29
Tecoma radicans, ... " 29
Fox Grape, ... " 29
Azalia viscosa, (white)..... " 29
Amphicarpæa monoica.... " 29
Utricularia, ... " 29
Pipe wort, ... " 29
Leucanthemum, ... " 30
Clover, ... " 30
Sparkle berry, (*Vaccinium*
    arboreum) .... " 30
Bay, (*Magnolia glauca*,)... " 30

Of the many valuable ornamental trees Lagerstrœmia Indica (Crepe Myrtle) has long ago been naturalized, attaining the height of 20 feet. Gardenia (Cape Jessamine) is hardy, and is used as a hedge or border in Oakdale Cemetery, where it fruits occasionally. The Japanese tea plants, distributed by the Agricultural Department from Washington. have been found hardy under the treatment of Mr. Donlan at Oakdale Cemetery, fruiting very freely. The Eucalyptus globulus was thoroughly tried by the gentleman above named, but with the utmost care the hardiest looking young tree was killed by the frost, and the experiment abandoned. The Sabal Palmetto, which has its northern limit at Smith's Island, has been successfully transplanted in three instances in Wilmington as an ornamental tree, much to the surprise of those who had formerly failed after great pains. Its transplantation seems to be as difficult as to transplant the pine. The banana during one favorable season fruited but did not come to full maturity. It is evident that the seasons are too short to expect success with it. The Ailantus is no longer a desirable shade tree, and has been generally made to give place to Elms and Oaks; and we are sorry to say that the Pride of China, one of the handsomest trees introduced among us, is losing popularity. A fine specimen of Negundo aceroides, from an accidental seed dropped in a private garden in the city, shows how kindly this beautiful tree would take to our soil if encouraged by the gardner. In our catalogue we have left out these, or most of them, and mention them here as an indication of the range of temperature.

This catalogue is arranged after the natural orders in Curtis' "Catalogue of Plants of North Carolina." The popular names have been retained as a matter of convenience—sometimes for curiosity.

We have added a list of plants found in the ballast along the river front, used to fill in the wharves of the Railroad Companies. The list is only a small portion of the plants found, but were all identified by Mr. McCarthy up to this time. These plants are from distant parts of the world, and it will be interesting to watch their history. They are mostly weeds, and from similar sources numbers of weeds have escaped into our fields and streets and by-ways. It will be noticed that several of these weeds, since Dr. Curtis' day, have become common in neglected streets and waste places, such as Viper's Bugloss (Echium) and Acanthospermum Xanthioides. Both plants are now stubborn weeds.

Species with names in heavy-face type are believed to be indigenous. Species with names in small capitals are not regarded as indigenous.

A numerical statement of the genera, species and varieties is given at the end of the catalogue.

9

CATALOGUE.

FLOWERING PLANTS.
DICOTYLEDONS.

RANUNCULACEÆ. (CROWFOOT FAMILY.)

Clematis crispa, Linn. (BLUE JESSAMINE. BLUE BELL.)
May and June until October. Flowers pale bluish purple. In three distinct crops.
    Virginiana, L. Rocky Point.   June. Flowers white.
    Viorna, L. Rocky Point. Dr. McRee.
May—August. Flowers purple.
Anemone Virginiana, L. (VIRGINIA ANEMONE.)
July—September. Flowers obscure white.
    Caroliniana Walt. Dr. McRee.   March. Flowers white.
Hepatica triloba, Chaix. Dr. McRee. Rocky Point.
February—March. Purple or white.
Thalictrum anemonoides, Michx. (RUE ANEMONE.)
April—May. Flowers white.
    Cornuti, L.—(MEADOW RUE.)   June - August. Flowers white.
Ranunculus abortivus, L. (SMOOTH CROWFOOT.)
March—April. Flowers yellow.

    BULBOSUS, L. (BUTTERCUPS.)   May. Flowers yellow.
    repens, L. (CREEPING CROWFOOT.)   March and April.
    repens var. nitidus, MUHL.   Flowers smaller than above.
    pusillus, Poir. (DWARF CROWFOOT.)
March and April. Flowers minute yellow.

    recurvatus, Poir. (*R. Nelsoni, Gray.*) (ROUGH CROWFOOT.)
April and May.

    parviflorus, L. (SMALL FLOWERED CROWFOOT.)
April and May. Flowers very small.

    palmatus, Ell. Rare.   April and May.
Aquilegia Canadensis, L. (COLUMBINE.) Mr. Norwood Giles, on the
Sound.   April and May. Flowers scarlet, yellow within.

MAGNOLIACEÆ. (MAGNOLIA FAMILY.)

Magnolia grandiflora, L. (MAGNOLIA.) Northern limit in Brunswick
county.   May. Flowers white.
    glauca, L. (SWEET BAY.)   May—June. Flowers white.
10

Umbrella, Lam. (UMBRELLA TREE.) May and June. Flowers white.
Liriodendron Tulipifera, L. (TULIP-TREE. POPLAR.)
May and June. Flowers greenish yellow.

ANONACEÆ. (CUSTARD APPLE FAMILY.)

Asimina parviflora, Dunal, (DWARF PAPAW. FETID SHRUB.)
March and April. Flowers greenish purple.

MENISPERMACEÆ. (MOONSEED FAMILY.)

Menispermum Canadense, L. (MOONSEED.) Dr. McRee. Rocky Point:
July. Flowers white.
Cocculus Carolinus, D. C. (REDBERRIED MOONSEED,)
June and August. Flowers white.

BERBERIDACEÆ. (BARBERRY FAMILY.)

Podophyllum peltatum, L. (MAY APPLE.) Not common.
April and May. Flowers greenish, moist shady places.

NELUMBIACEÆ. (NELUMBO FAMILY.)

Nelumbium luteum, Willd. (DUCK ACORN. WATER CHINQUAPIN.)
Is rarely found in Waccamaw Lake, 30 miles from Wilmington.
Found in Neuse river by G. McCarthy, July, 1886.
July. Flowers large yellow.

CABOMBACEÆ. (WATER-SHIELD FAMILY.)

Cabomba Caroliniana, Gray. (*Nectris aquatica Nutt.*)
Ditches in Potter's rice field. June and August. Flowers white.
Brasenia peltata, Pursh. (*Hydropeltis purpurea, Michx.*) (WATER-SHIELD)
July. Flowers dull purple, under surface of leaves with gelati-
nous viscid coating.

NYMPH.EACEÆ. (WATER-LILY FAMILY.)

Nymphæa odorata, Ait. (WHITE POND-LILY.)   May and June.
Nuphar advena, Ait. (YELLOW WATER-LILY. BONNETS. SPATTER DOCK.)
sagittæfolia, Pursh. (ALLIGATOR BONNETS.)
June—August. Flowers yellow. Found on the margins of the
Cape Fear N. E. River and Smith's Creek. Plentiful enough
to make their locality a favorite place for fishermen.

SARRACENIACE.E. (PITCHER-PLANT FAMILY.)

Sarracenia purpurea, L. (Pitcher-plant.)
April and May. Flowers purple.
Varieties with erect leaves.
Sarracenia rubra, Walt. (ED-FLOWERED TRUMPET.)
Found in the neighborhood of Rush's. See Map.
flava, L. (Biscuit. Yellow Trumpets. Watches.)   March,
variolaris, Michx. (*S. lacunosa of Bartram.*) Spotted.
Trumpet-leaf, 30 miles from Wilmington. Scarce.

PAPAVERACEÆ. (POPPY FAMILY.)

Argemone Mexicana, **L.** (Mexican Poppy.) Introduced.
April and May. Flowers yellow and white. Yellow flowers two
or three weeks before white.
var. alba.
Sanguinaria Canadensis, L. (Blood-root. Puccoon-root.)
Not common. March. Flowers white.
Papaver dubium, (*P. rheas,*) Dr. McKee, Rocky Point.
Cultivated fields. Adventive.

FUMARIACE.E. (FUMITORY FAMILY.)

Corydalis aurea, Willd.   March and April. Flowers yellow.
micrantha, Dr. A. Havard. Smithville.

CRUCIFER.E. (MUSTARD FAMILY.)

Cardamine rhomboidea, D. C. (Spring Cress.) Dr. McKee.
Rocky Point. April and May. Flowers white.
hirsuta, L. (Bitter Cress)
Cakile maratima, Scop. (Sea Kale.) On the sea beach.
May—August. Flowers pale purple.
Sisymbrium Thaliana, **Gaud.** (Mouse-ear Cress.)
March and April. Flowers White.
canescens, Nutt. (Tansy Mustard.)
March and April. Flowers small greenish white.
officinale, Scop. (Hedge Mustard.) Introduced.
May and June. Flowers pale yellow.
Draba verna, L. (Whitlow Grass.)   Flowers small white.
Caroliniana, Walt.   February—April. Flowers white.
Senebiera pinnatifida, D. C. (Wart Cress. Swine Cress.)
March—May.
Lepidium Virginicum, L. (Wild Pepper-grass.)
March—June. Very common.
Capsella Bursa-pastoris, **Mœnch.** (Shepherd's Purse.) Introduced.
March—June. Flowers white.

CAPPARIDACE.E. (CAPER FAMILY.)

Gynandropsis pentaphylla, D. C.   May—August. Flowers white.
    (*Cleome pentaphylla, Linn, and Barton, Pursh, Nuttall anp
    Elliott.*) Waste grounds. Across Wilmington ferry, Cape
    Fear river. Dr. McKee.

VIOLACE.E. (VIOLET FAMILY.)

Viola cucullata, Ait. (BLUE VIOLET.)
               February—May. Flowers blue-white.
palmata, L. (HAND-LEAF VIOLET.) Not common.
               March—May. Flowers purple or blue.
villosa, Walt. (HAIRY VIOLET.)   Flowers pale blue.
lanceolata, L. LANCE-LEAVED VIOLET.)
               February—May. Floweis white.
pedata, L. (BIRD-FOOT VIOLET.)
               February—May. Flowers large, deep blue or purple.
primulæfolia, L. (PRIMROSE-LEAVED VIOLET.)
               Flowers white—small.
blanda, Willd. (SWEET WHITE VIOLET.)
               April—May. Flowers white.
tricolor, L. Var. arvensis D. C. (WILD PANSY.)
               Not very common.

CISTACEÆ. (ROCK-ROSE FAMILY.)

Helianthemum Canadense, Michx. (FROST-WEED.)   April.
    [*Cistus Canadensis, Walt.*]

Carolinianum, Michx. (ROCK-ROSE.) March—April. Flowers yellow.
    [*Cistus Carolinianus, Walt.*]

corymbosum, Michx.   April.
    [*Cistus corymbosus, Pair.*]
Lechea major, Michx. (PIN-WEED.)   July—August. Greenish flowers.
minor, Lam. Dry sandy soil.   July and August.

DROSERACEÆ. (SUNDEW FAMILY.)

Dionæa muscipula, Ellis. (VENUS FLY-TRAP.)*
    June. Flowers white, blacken in drying. Seed drop in July-
    10th to 20th.
    Found abundantly in a radius of a half to three quarters of a mile
    from 2d toll house and other places. See Map.

---

*The original description of the habit of Dionæa, written by Dr. Curtis for his first "Catalogue of Plants Growing Spontaneously Around Wilmington," in 1834, is reproduced, as it stands a most excellent, and perhaps the best sketch yet made.

"DIONÆA MUSCIPULA.—This plant is found as far north as Newbern, N. C., and from the mouth of Cape Fear river nearly to Fayetteville. Elliott says, on the authority of Gen. Pinckney, that it grows along the lower branches of the Santee in South Carolina. Dr. Bachman has received it from Georgetown, *S. C.; and Mr. Audubon informed me, with the plant before us, that he has seen it in Florida of enormous size. I think it not impossible, therefore, that it inhabits the savannahs, more or less abundantly, from the latter place to Newbern.† It is found in great abundance for many miles around Wilmington, in every direction. I venture a short notice of this interesting plant, as I am not aware that any popular description of it has been published in this country.

The leaf, which is the only curious part, springs from the root, spreading upon the ground, or at a little elevation above it. It is composed of a petiole or stem with broad margins, like the leaf of an orange tree, two to four inches long, which at the end suddenly expands into a thick and somewhat rigid leaf, the two sides of which are semi-circular, about two-thirds of an inch across, and fringed around their edges with somewhat rigid ciliæ or long hairs like eye-lashes. It is very aptly compared to two upper eye-lids joined at their bases. Each side of the leaf is a little concave on the inner side, where are placed three delicate, hair-like organs, in such an order that an insect can hardly traverse it without interfering with one of them, when the two sides suddenly collapse and enclose the prey with a force surpassing an insect's efforts to escape. The fringe or hairs of the opposite sides of the leaf interlace, like the fingers of the two hands clasped together. The sensitiveness resides only in these hair-like processes on the inside, as the leaf may be touched or pressed in any other part without sensible effects. The little prisoner is not crushed and suddenly destroyed, as is sometimes supposed, for I have often liberated captive flies and spiders which sped away as fast as fear or joy could hasten them. At other times I have found them enveloped in a fluid of mucilaginous consistence, which seems to act as a solvent, the insects being more or less consumed in it. This circumstance has suggested the possibility of their being made subservient to the nourishment of the plant, through an apparatus of absorbent vessels in the leaves. But as I have not examined sufficiently to pronounce on the universality of this result, it will require further observation and experiment on the spot to ascertain its nature and importance. It is not to be supposed, however, that such a food is necessary to the existence of the plant, but like compost, may increase its growth and vigor. But however obscure and uncertain may be the final purpose of such a singular organization, if it were a problem to construct a plant with reference to entrapping insects, I cannot conceive of a form and organization better adapted to secure that end than are found in the Dionæa Muscipula. I therefore deem it no credulous inference, that its leaves are constructed for that specific object, whether insects subserve to the purpose of nourishment to the plant or not. It is no objection to this view that they are subject to blind accident, and sometimes close upon straws as well as insects. It would be a curious vegetable, indeed, that had a faculty of distinguishing bodies, and recoiled at the touch of one, while it quietly submitted to violence from another. Such capricious sensitiveness is not a property of the vegetable kingdom. The spider's net is spread to ensnare flies, yet it catches whatever falls upon it ; and the ant lion is roused from her hiding place by the fall of a pebble ; so much are insects also, subject to the blindness of

---

*I find upon diligent inquiry that it is not to be found at Georgetown but near Bucksville, S. C., about 70 miles from Wilmington, and is very scarce there.   T. F. W.

†The above was written in 1834, but has not since been verified. Dionæa is quite as local as at first found to be, Wilmington being the centre of its most abundant growth.

accident. Therefore the web of the one, and the pitfall of the other are not designed to catch insects ! Nor is it in point to refer to other plants of entirely different structure and habit which sometimes entangle and imprison insects. As well might we reason against a spider's web, because of a fly drowned in a honey pot ; or against a steel trap because some poor animal has lost its life in a cider barrel."

The extent of the distribution of the Dionæa has interested many botanists, and we append some observations on that subject. Although drainage of the savanna land near Wilmington, and frequent accidental fires and fires purposely employed to clear off the savannas to secure grazing for cattle, after the wasteful method handed down by the aborigines, have lessened the numbers of the plant immediately about the town, still some good specimens are occasionally found within the unsettled limits. The distribution of Dionæa is co-extensive with the savanna land, which, it will be safe to say, is as much as one-fifth of the area of this county (New Hanover,) and the adjacent one's of Pender and Brunswick counties. The ravages of fire, and unusual cold seasons and the rapacity of the collectors who supply the trade, have scarcely made perceptible inroads on the plant.

The future of the Dionæa does not incline us to believe that it will be soon exterminated. Land is very abundant, and but few savannas have more than a superficial coating of organic matter overlaying a bed of almost sterile sand, therefore the temptation to cultivate such land will be small, even after the country is thickly settled. It will be a great many years before the population of farmers will become dense enough to regard savanna land as worth much, even for the grazing of stunted cattle, and as the very sensible plan of caring for cattle (and the better breeds are being rapidly introduced) upon the economic basis of good shelter, pure water, and wholesome food becomes an established fact, there will be no temptation to do as wasteful a thing as to burn off a savanna and risk a forest of good trees, to keep a few stunted cattle from starving. The Dionæa is undoubtedly the most remarkable of the very local plants known to botanists, and it will be good news to the friends of science to know its probable future.

Drosera filiformis, Raf. (THREAD-LEAVED SUNDEW.)
April. Flowers bright purple.

longifolia, L. (LONG-LEAVED SUNDEW.)
May and June. Flowers white.

rotundifolia, L. (ROUND-LEAVED SUNDEW.)
. May and June. Flowers white.

brevifolia, L. (SHORT-LEAVED SUNDEW.) March. Flowers white:

PARNASSIACEÆ

Parnassia Caroliniana, Michx. (GRASS of PARNASSUS.)
October and November. Corolla white with impressed greenish veins.

HYPERICACEÆ. (ST. JOHN'S WORT FAMILY.)

Ascyrum Crux-Andreæ, L. (ST. PETER'S WORT.)
June—September. Flowers yellow.

stans, Michx.
July—September. Flowers yellow.

*Hypericum angulosum, Michx.   June—August.   Flowers almost orange.
   corymbosum, Muhl.   July.   Petals black dotted.
   fasciculatum, Lam.   June—August.   Petals yellow.
   nudiflorum, Michx.   July—August.
   mutilum, L.   (*H. parvifolium, Willd. Pursh and Elliott.*)
                  June—August.   Flowers minute orange colored.
   *rosmarinifolium.   (*H. cistifolium, myrtifolium.*)
   *simplex.   (*H. setosum.*)
   †Virginicum.   (*H. corymbosum.*)
   Canadense, L.   June—October.   Flowers copper yellow.
   Sarothra, Michx.   (GROUND PINE.)
         [*Sarothra gentianoides, L.*] June—August.   Flowers small yellow.

PORTULACACEÆ. (PURSLANE FAMILY.)

Claytonia Virginica, L.   (SPRING BEAUTY.)
               April—May.   Flowers rose colored, veiny; not common.
Portulaca OLERACEA, L.   (PURSLANE.)   Common in streets and cultivated
      fields.   Introduced.   May—July.   Flowers yellow.
Sesuvium pentandrum, Ell.   (SEA PURSLANE.)   Saline marshes.
                                                      May—November.

   portulacastrum, L.   May—November.

CARYOPHYLLACEÆ. (PINK FAMILY.)

Arenaria Canadensis* (*diffusa?*) (*A. rubra*, found near Sound. Dr. McRee.)
   diffusa, Ell.   April—June.   Flowers white.
      SERPYLLIFOLIA, L.   (THYME-LEAVED SANDWORT.)   Introduced.
                                              June.   Petals white.
Cerastium VISCOSUM, L.   (*C. hirsutum, Muhl., Ell., &c.*)   Introduced.
                                    February—May.   Flowers white.
      VULGATUM, L.   (MOUSE-EAR CHICKWEED.)   Introduced.
                                    February—May.   Flowers white.
Mollugo verticillata, L.   (INDIAN CHICKWEED.)
                            May—August.   Flowers small white.
Alsine squarrosa, Fenzl.   (BARRENS SANDWORT.)   Dry sand hills.
                            April—May.   Flowers white.
Saponaria OFFICINALIS, L.   (SOAPWORT. BOUNCING BET.)   Introduced.
         Waste grounds.   July—August.   Flowers white or rose-color.
Silene Antirrhina, L.   (CATCHFLY.)   July.   Flowers small red.
   Virginica, L.   (INDIAN PINK.)   June.   Flowers large red.
   Pennsylvanica, Michx.   Lilliput on Cape Fear River. (Dr. McRee.)
                            April.   Petals white or rose-color.

---

*Not given in Curtis' Catalogue of Indigenous Plants. See page 12.

†Synonym for *corymbosum* given above.

   10

Spergula ARVENSIS, L. (PINE CHEAT. CORN SPURREY.)
May—August. Flowers white.
Spergularia rubra, Pers. (SAND SPURREY.) Sea-coast.
May—October. Flowers red or rose.
Stellaria MEDIA, **Smith.** (CHICKWEED.) Introduced.
December—April. Flowers white.
uniflora, Walt.
Paronychia herniarioides, Nutt. July—October. Flowers minute.
Anychia dichotma, Michx. July—August. Flowers greenish.
Stipulicida setacea, Michx. White sand hills. Common.
April—June. Flowers white.
Sagina Elliottii, Fenzl. (BARRENS SANDWORT.) April—June.

MALVACEÆ. (MALLOW FAMILY.)

Hibiscus Moscheutos, L. (SWAMP MALLOW.)
July. Flowers white or pale rose, crimson centre.
Hibiscus aculeatus, Walt. (*H. Scaber, Michx.*)
July. Flowers white with purple centres.
militaris, Cav. (*H. Virginicus, Walt.*) (ROSE-MALLOW.)
Causeway and Little Bridge. July—August. Flowers rose.
Malva ROTUNDIFOLIA, L. (MALLOW.) Introduced. Common in waste grounds.
May—August. Petals pale pink.
Sida SPINOSA, L. Common about settlements. Introduced. See Abutilon.
July—August. Petals yellow.
[*S. Abutilon, Avicennæ, Gærtn.*]
Elliottii, T. and G. July—October. Flowers orange yellow.
Abutilon AVICENNÆ, **Gaert.** (VELVET LEAF AMERICAN HEMP.) Introduced.
June—July. Flowers yellow.
Kosteletzkya Virginica, Presl.

TILIACEÆ. (LINDEN FAMILY.)

Tilia pubescens, Ait. (LINN OR LIME TREE.) June. Flowers cream color.
heterophylla, Vent. (White Linn.)
Americana, L. Dr. McRee. Rocky run.
[*Tilia glabra, Vent*]

CAMELLIACEÆ. (CAMELLIA FAMILY.)

Gordonia Lasianthus, L. (LOBLOLLY BAY. BLACK LAUREL.)
July—August. Flowers white.
Stuartia Virginica, Cav. Scarce.
April—May. Flowers white, stamens purple.
Found eight miles W. from Wilmington by Mr. Wm. Watters.

MELIACEÆ.

Melia AZEDARACH, L. (CHINA TREE. PRIDE OF INDIA.)
May—June. Flowers lilac.

LINACEÆ. (FLAX FAMILY.)

Linum Virginianum, L. (WILD FLAX.) July. Flowers yellow.
Bootii, Planch. Dry pine woods. July. Flowers sulphur yellow.

OXALIDACEÆ. (WOOD-SORREL FAMILY.)

Oxalis stricta, L. (YELLOW WOOD SORRELL.) April. Flowers yellow.
violacea, L. (PURPLE WOOD SORRELL.) Dr. McRee, Rocky Point.
May—June. Flowers nodding, purple.

ZYGOPHYLLACEÆ. (BEAN CAPER FAMILY.)

Tribulus CISTOIDES, L. Waste grounds. Dr. McRee. Flowers large yellow.
TERRESTRIS, L. Ballast. Introduced from South Russia.

GERANIACEÆ. (GERANIUM FAMILY.)

Geranium Carolinianum, L. March—April. Flowers pale purple.
maculatum, L. (CRANE'S BILL.) Dr. McRee.
April—May. Flowers purple.

BALSAMINACEÆ. (BALSAM FAMILY.)

Impatiens fulva, Nutt. (JEWELL WEED.) Causeway, rice fields, Little
Bridge. July—September. Flowers deep orange.
pallida, Nutt. (TOUCH-ME-NOT.) Gerald McCarthy.

RUTACEÆ. (RUE FAMILY.)

Zanthoxylum, Carolinianum, Lam. (PRICKLY ASH. TOOTHACHE TREE.)
[*Z. clava-Herculis, Linn.*] June. Flowers small greenish.
Ptelea trifoliata, L. (HOP-TREE. WAFER-ASH.) Wrightsville Sound.
May—June. Flowers greenish.

mollis, M. A. C. May—June. Flowers greenish.

ANACARDIACEÆ. (CASHEW FAMILY.)

Rhus copallina, L. (COMMON SUMACK.)   July.   Flowers greenish.
    Toxicodendron, L. (POISON OAK.)   July.   Flowers small·
    radicans, L. (POISON VINE.)   July.   Flowers greenish.
Rhus venenata, D. C. (POISON SUMACH.)
                         July.   Flowers very small—green.
    pumila, Michx. (DWARF SUMACH.)   Pine woods.

VITACEÆ. (VINE FAMILY.)

Vitis æstivalis, Michx. (SUMMER GRAPE.)
                      June.   Grape deep blue, very austere.
    Labrusca, L. (FOX GRAPE.)
                  May—June.   Grape purple or whitish, pleasant.
    vulpina, L. (MUSCADINE.   BULLACE.)
                  June.   Grape purple, pleasant.
    cordifolia, Michx. (FROST GRAPE.) May—June.   Grape black, acid.
    bipinnata, Torrey & Gray.   June—July.   Grape small, black.
Ampelopsis quinquefolia, Michx. (VIRGINIA CREEPER.)
                June.   Flowers inconspicuous, greenish.

RHAMNACEÆ. (BUCKTHORN FAMILY.)

Ceanothus Americanus, L. (RED-ROOT.   JERSEY TEA.)
                    July.   Flowers white.
Berchemia volubilis, D. C. (RATTAN.   SUPPLE JACK.)
                 June.   Flowers small, greenish.
Sageretia Michauxii, Brongn.   Sea coast.   September.   Flowers white.
Frangula Caroliniana, Gray.   June.

CELASTRACEÆ. (STAFF-TREE FAMILY.)

Euonymus Americanus, L. (STRAWBERRY BUSH.   BURSTING HEART.
    FISH-WOOD.)   May—June.   Flowers greenish.

STAPHYLLACE.E. (BLADDER-NUT FAMILY.)

Staphylea TRIFOLIA, L. (BLADDER NUT.)   Introduced.
                  May.   Flowers white.

SAPINDACEÆ. (SOAP-BERRY FAMILY.)

Æsculus Pavia, T. (RED BUCKEYE.   HORSE-CHESTNUT.)
                March—May.   Flowers red.

ACERACEÆ. (MAPLE FAMILY.)

Acer rubrum, L. (RED MAPLE.) February—March. Flowers and fruit red.
saccharinum, Wang. (SUGAR MAPLE.) Rock spring. Rocky Point.
April and May.
Negundo aceroides, Mœnch. (ASH-LEAVED MAPLE.)
Rocky Point and in Wilmington. March and April.

POLYGALACEÆ. (MILKWORT FAMILY.)

Polygala cymosa, Walt. Moist savannahs. Common.
July. Flowers yellow, turning dark green in drying.
cruciata, L. July—October. Flowers pale rose color.
incarnata, L. Moist savannahs. Common.
June—August. Flowers purple.
lutea, L. (BATCHELOR'S BUTTON.) Savannahs. Abundantly common.
June—August. Flowers orange yellow.
ramosa, Ell. July—September. Flowers yellow.
fastigiata, Nutt. Not as common as others.
July—October. Flowers small, bright rose color.
setacea, Michx. March—July. Flowers pale rose color or whitish.
brevifolia, Nutt. July—October. Flowers reddish purple.
grandiflora, Walt. July—September. Flowers bright purple.
Verticillata, L. June—August. Flowers greenish white.

LEGUMINOSÆ. (PULSE FAMILY.)

Amorpha fruticosa, L. (INDIGO BUSH.) May—June. Flowers blue.
herbacea, Walt. (A. PUBESCENS WILLD.)
June—July. Flowers blue or white.
Amphicarpæa monoica, Nutt. (PEA-VINE.)
August—Septempter. Flowers white or purplish.
Apios tuberosa, Mœnch. (GROUND NUT.)
This common name must not be confounded with "ground nut,"
the local name for ARACHIS HYPOGEÆ, the introduced African
"pea-nut" of commerce.
July—August. Seeds black. Flowers brownish purple.
Astragalus glaber, Michx. April. Flowers white.
Baptisia lanceolata, Ell. April—May. Flowers yellow.
villosa, Ell. May. Plants turns black in drying.
alba, R. Br. April. Flowers white.
tinctoria, R. Br. (Wild Indigo.) May—June. Flowers yellow.
Cassia Chamæcrista, L.
July—August. Flowers yellow, petals often purple at the base.
Marylandica, L. (WILD SENNA.) August. Flowers yellow.
occidentalis, L. July. Flowers large, yellow.
obtusifolia, L. (C. Tora Linn.) July—October. Flowers yellow.
nictitans, L. (WILD SENSITIVE PLANT.) July. Flowers pale yellow.

Cercis Canadensis, L. (RED-BUD.)   March.   Flowers rose colored.
Crotalaria sagittalis, L. (RATTLE-BOX.)   June—July.   Flowers yellow.
    ovalis, Pursh.   April—June.   Flowers showy, yellow.
    Purshii, D. C.   May—July.   Flowers yellow.
Lupinus diffusus, Nutt.   April—May.   Flowers blue.
    perennis, L. (LUPINE.)   ·
      April—May.   Flowers purplish or purplish blue.
    villosus, Willd.   April—June.   Flowers violet and roseate above.
Trifolium ARVENSE L. (RABBIT-FOOT CLOVER.)
      July—August.   Flowers pale red.
    Carolinianum, Michx. (CAROLINA CLOVER.)   Dr. McRee, Rocky
      Point.   Introduced.   March—May.   Flowers white or purplish.
    pratense, L. (RED CLOVER.) Introduced.   All summer.
    hybridum, L.* Gerald McCarthy.
    repens, L. (WHITE CLOVER.)   May—September.
    reflexum, L. (BUFFALO CLOVER.)   April—June.   Flowers rose red.
    PROCUMBENS, L. (YELLOW CLOVER.)   Introduced.   June—July.
Medicago LUPULINA, L. (HOP, MEDICK.) Common in grass plats. Introduced.
      May—October.   Flowers small, yellow.
    denticulata.* G. McCarthy.
Psoralea melilotoides, Michx.   May—July.   Flowers violet.
    canescens, Michx. (BUCK ROOT.)   May—July.   Flowers blue.
    lupinellus, Michx.   May—June.   Flowers.
Robinia PSEUD-ACACIA, L. (WHITE LOCUST.)
      Introduced as shade trees from the mountains.   March—April.
    hispida, L. (ROSE LOCUST.)   April—May.   Flowers rose.
      var. nana, Ell. PINE WOODS.
Wistaria frutescens, D. C. (VIRGIN'S BOWER.)
      (*Thyrsanthus frutescens*, Ell.)   April—May.   Flowers lilac.
†**Tephrosia Virginiana, Pers.** (RABBIT PEA.)
      July.   Banner white, heel rose colored, wings red.
    hispidula, Pursh.   May—August.   Flowers reddish purple.
    ambigua, M. A. C.   June—July.   Flowers white and purple.
    spicata, T. & G.   June—July.   Flowers large, white and purple.
Indigofera Caroliniana, Walt. (CAROLINA INDIGO.)
      July—September.   Flowers reddish brown.
Lathyrus paluster, Linn.   June—July.   Flowers blue and purple.
      ‡**var. myrtifolium,** Gray.
Æschynomene hispida, Willd. (*Hedysarum* —— ?)
      August.   Flowers small, yellow.
Zornia tetraphylla, Michx.   June—August.   Flowers deep yellow.

---

*Not given in Curtis Catalogue.

†Tephrosia is substituted as the name of this genus, following Curtis' Catalogue of Indigenous Plants, p. 17.

‡Watson's Index, p. 230.

Styloshanthes elatior, Swartz. (PENCIL FLOWER.)
July—August. Flowers yellow.

Lespedeza capitata, Michx. (BUSH CLOVER.) August—September.
repens, T. & G. Common. August—September.
procumbens.* G. McCarthy.
violacea, Persoon. July—August. Flowers.
" **var. sessiliflora, Persoon.** July—August.
hirta, Ell. August—September. Flowers reddish white.
stuvei, Nutt.

†**Desmodium nudiflorum, DeCandolle.** (*Hedysarum nudifolium*, Linn.)
August. Flowers small, purple.
cuspidatum, T. & G. August. Flowers large, purple.
viridiflorum, Beck. (*H. viridifolium, Elliott.*)
August. Flowers yellow, green when dry.
rotundifolium, DeCan. (*H. rotundifolium, Michx.*)
August. Flowers purple.
ochroleucum, M. A. C. August. Flowers yellow.
Dillenii, Darlington. (*Hedysarum Marilandicum, Willd.*)
July. Flowers purple.
glabellum, DeCan. August—September.
paniculatum, DeCan. July—August. Flowers purple.
strictum, DeCan. August. Flowers small, purple.
Marilandicum, Boott. August. Flowers violet purple.
rigidum, DeCan. August. Flowers violet purple.
lineatum, DeCan. Flowers and legume small.

Rhynchosia tomentosa, T. & G. (*Glycine tomentosa, Linn.*)
Flowers yellow.
" **var. monophylla, T. & G.**
" **var. volubilis, T. & G.**
" **var. erecta, T. & G.**

Gleditschia triacanthos, L. (HONEY LOCUST.)
June. Flowers small, green.

Clitoria Mariana, Linn. (*Vexillaria Mariana Eaton.*)
July—August. Flowers pale purple.

Phaseolus perennis, Walt. (WILD BEAN.)
July—August. Flowers purple and violet.
diversifolius, Pers. August—October. Flowers purplish.
helvolus, L. August—September. Flowers purplish.

Erythrina herbacea, L. April—June. Flowers deep scarlet.
Galactia pilosa, Nuttall. June—September. Flowers roseate.
var. Macræi,* M. A. C. (Watson's Index, 221)
glabella, Michx.

---

*Not given in Curtis' Catalogue.

†Generic name Desmodium substituted for Hedysarum. See Cat. Indig. Plants, p. 19.

Schrankia angustata, T. & G. (SENSITIVE PLANT.)
                    May—July.   Flowers purplish.
Arachis HYPOGÆA, **Willd.** (PEA-NUT. GROUND-NUT.)   .
        Introduced and cultivated as a crop plant.
                    May—August.   Flowers yellow.
Vicia SATIVA, **L.** (VETCH. TARE.) Introduced—cultivated grounds.
                    April—May.   Flowers pale purple.
        HIRSUTA, **Koch.** Introduced.   April—May. Flowers blueish white.
        TETRASPERMA, **Loisel.**
Melilotus OFFICINALIS, **Willd.** (YELLOW MELILOT.)
        Naturalized about old clearings.
        ALBA, **Lam.** (WHITE MELILOT.) Naturalized about old clearings.
Petalastemon corymbosum, Michx.
                    September—October.   Flowers white.
Centrosema Virginiana, Benth. Dry soil.   June—September.

ROSACEÆ. (ROSE FAMILY.)

Agrimonia Eupatoria, L. (FEVERFEW.)   July.   Flowers yellow.
    parviflora, Ait.   Aug 1st. Flowers (petals) yellow.
Fragaria Virginiana, Ehrhenberg. (STRAWBERRY.)
                    January—May.   Flowers white.
        INDICA, **Ait.** (INDIAN STRAWBERRY.) April—May.   Flowers white.
Geum album, Gmelin. (AVENS.)   July. Flowers small, white.
Potentilla NORVEGICA, **L.** Rare. Introduced.
                    July—September.   Flowers pale yellow.
    Canadensis, L. (FIVE-FINGER.) Very common.
                    April—August.   Flowers yellow.
Rosa Carolina, L. (SWAMP ROSE.) Common, most in wet grounds.
                    May—June.
    lucida, Ehrh. (WILD OR DWARF ROSE.) Common in dry woods.
                    May—July.
        RUBIGINOSA, **L.** (SWEET-BRIER.) Near settlements. Introduced.
                    May. Flowers orange-red.
        LÆVIGATA, **Michx.** (CHEROKEE ROSE.) Meares' Bluff.
                    April.   Flowers white.
Rubus villosus, Ait. (HIGH BLACKBERRY.)
                    February—April.   Flowers white.
    cuneifolius, Pursh. (LOW BLACKBERRY.)
        Common on old fields and roadsides.
                    February—March.   Flowers white.
    trivialis, Michx. (DEWBERRY.) February—April.   Flowers white.
Prunus Americana, Marsh. (RED PLUM.) March—April. Flowers white.
    Chicasa, Michx. (CHICASAW PLUM.)   April.   Flowers white.

Prunus Caroliniana, Ait. (MOCK ORANGE.) Dr. McRee.
　　Seacoast of Brunswick county. Introduced as an ornamental
　　shade tree. Fruit very poisonous.
Spiræa tomentosa, L. (HARDHACK.)
　　　　　　　　　June—July. Flowers small, stamens purple.
　opulifolia, L. (HARDHACK.) Ger. McCarthy.
Cratægus spathulata, Michx. (NARROW-LEAVED THORN.)
　　　　　　　　　April—May. Flowers small, white.
　glandulosa, Michx. (HÁIRY THORN.)
　flava, Ait. (SUMMER HAW.)　　　　April—May. Flowers white.
　parvifolia, Ait. (DWARF THORN.)
　apiifolia, Michx. (PARSLEY-LEAVED HAW.)
　　　　　　　　　March—April. Flowers large, white or roseate.
　crus-galli, L. (COCK'S-SPUR THORN.)
Pyrus angustifolia, Ait. (NARROW-LEAVED CRAB.)
　　　　　　　　　　　　March. Flowers rose purple.
　arbutifolia, L. (CHOKEBERRY.)　　March—May. Flowers white.
Amelanchier Canadensis, L. (SERVICE TREE.)
　　　　　　　　　March—May. Flowers white.

CALYCANTHACEÆ. (CAROLINA ALL.-SPICE FAMILY.)

Calycanthus FLORIDUS, **L.** (SWEET SHRUB.) Introduced.
　　　　　　　　　April and May. Flowers brown.

MELASTOMACEÆ. (MELASTOMA FAMILY.)

Rhexia Mariana, L.　　　　　June—September. Flowers purple.
　var. lanceolata. (*R. angustifolia, Nutt.*)
　　　　　　　　　June—August. Flowers white or pale purple.
　***Virginica, L.** (MEADOW BEAUTY.)
　　　　　　　　　July and August. Flowers bright purple.
　glabella, Michx. (DEER-GRASS.)
　　　　　　　　　June—August. Flowers pale purple.
　lutea, Walt.　　　　　　June—August. Flowers yellow.
　ciliosa, Michx.　　　　　June—August. Flowers purple.

LYTHRACEÆ. (LOOSESTRIFE FAMILY.)

Lythrum alatum, Pursh. (LOOSESTRIFE.)　June—July. Flowers purple.
　lineare, L.　　　　　　　June. Flowers nearly white.
Nesæa verticillata, H. B. & K. (SWAMP LOOSESTRIFE.)
　　　　　　　　　·July—August. Flowers purple.
Ammannia humilis, Michx. (var. ramosior, Michx.)　Petals purplish.

---

*Curtis does not corroborate this in Cat. Indig. Plants.
　I I

ONAGRACEÆ. (EVENING-PRIMROSE FAMILY.)

Gaura angustifolia, Michx. Dry old fields and sandy places.
Flowers white.

Œnothera biennis, L. (EVENING PRIMROSE.)
Green's (Kidder's) lower rice field. Common mostly in planta-
tions. Flowers yellow.

fruticosa, L. (SUNDROPS.) River banks. June. Flowers yellow.

sinuata, L. Drift sand, near coast.
May—September. Flowers small, axillary.

var. humifusa. Sea beach.

riparia, Nutt. Swamps and river banks. June—July.

Jussiæa decurrens, D. C. (*Ludwigia decurrens.*) Sides of ditches. July.

Ludwigia alternifolia, L. (SEED-BOX.) (*L. macrocarpa Michx.*)
August. Flowers yellow.

capitata, Michx. July—August. Flowers yellow.

linearis, Walter. July—September. Flowers yellow.

pilosa, Walter. (*L. Mollis Michx.*)
July—September. Flowers yellow.

virgata, Michx. July—August. Flowers yellow.

hirtella, Rob. Pine woods. August.

palustris, Elliott. (WATER PURSLANE.) June—September.

linifolia, Poir. Ditches and swamps. July—September.

natans, Elliott. July—September.

arcuata, Walter. Ditches and margin of ponds. July.

sphærocarpa, Elliott. Rare. July—September.

Proserpinaca palustris, L. (MERMAID WEED.)
June—August. Flowers greenish.

pectinacea, Lam. July—August.

Myriophyllum verticillatum, L. (WATER MILFOIL.)
July. Flowers small, green axilla.

heterophyllum, Michx. July

CACTACEÆ. (CACTUS FAMILY.)

Opuntia vulgaris, Mill. (PRICKLY PEAR.) (*Cactus opuntia.*)
May—June. Flowers yellow.

PASSIFLORACEÆ. (PASSION FLOWER FAMILY.)

Passiflora incarnata, L. (MAY-POP. PASSION FLOWER.)
June—July. Flowers purple and white.

lutea, L. June—July. Flowers small, greenish yellow.

CUCURBITACEÆ. (GOURD FAMILY.)

Lagenaria VULGARIS, **Sering.** (GOURD.) About settlements.
May—June. Flowers yellow.
Sicyos angulatus, L. (ONE-SEEDED CUCUMBER.) G. McCarthy.
Melothria pendula, L. In rice fields and elsewhere.
May—August. Flowers small, yellow.

CRASSULACEÆ. (ORPINE FAMILY.)

Penthorum sedoides, L. Ditch stone crop. July—September.

SAXIFRAGACEÆ. (SAXIFRAGE FAMILY.)

Saxifraga Virginiensis, Pursh. (EARLY SAXIFRAGE.)
April—May. Flowers white.
Heuchera Americana, L. (ALUM-ROOT.) April—May. Flowers white.
Itea Virginica, L. May—June. Flowers white.
Decumaria barbara, L. (*D. sarmentoso, Ell.*)
May—June. Flowers odorous, white.

HAMAMELACEÆ. (WITCH-HAZEL FAMILY.)

Hamamelis Virginica, L. (WITCH-HAZEL.) Winter. Flowers yellow.
Fothergilla alnifolia, L. (DWARF ALDER.) March—April. Flowers white.
Liquidambar Styraciflua, L. (SWEET-GUM.) · May.

UMBELLIFERÆ. (PARSLEY FAMILY.)

Hydrocotyle repanda, Pers. June—August.
umbellata, L. (WATER GRASS.) May—July.
ranunculoides, L. July—August.
interrupta, Muhl. June.
Cicuta maculata, L. (WATER HEMLOCK. WILD PARSNIP.)
July. Flowers white.
Daucus pusillus, Michx. (DWARF CARROT.) Smithville.
June. Flowers white or yellowish.
Eryngium yuccæfolium, Michx. (BUTTON SNAKEROOT.) (*E. aquaticum, Linn.*)
June. Flowers blue.
Virginianum, Lam. July. Flowers blue.
virgatum, Lam. August. Flowers blue.
præaltum, Gray. (*E. Virginianum, Ell.*) August. Flowers white.
Archemora rigida, D. C. (COWBANE. PIG POTATOE.) (*Sium denticulatum.*)
August—September. Flowers white.
ternata, Nutt. November. Flowers white.

Sanicula Canadensis, L. (SANICLE.)   May.
    Marylandica, L.   May.
Discopleura capillacea, D. C. (BISHOP WEED.) (*Sison capillacens.*)
                                               June—July. Flowers white.
    costata.* Gerald McCarthy.
Tiedemannia teretifolia, D. C. (WATER DROP-WORT.) (*Oenanthe filiformis, Walt.*)
    mis, Walt.)   August. Flowers white.
Crantzia lineata, Nutt. Muddy banks of rivers.
                                      July. Flowers small whitish.
Pastinaca SATIVA, **L.** (PARSNIP.) About settlements.
Leptocaulis divaricatus, D. C. Dry sandy soil.   April. Flowers white.
Thaspium barbinode, Nutt. River banks.
                          May—June. Flowers pale yellow.
    aureum, Nutt. (MEADOW PARSNIP.) Rocky Point—rich soil.
                              May. Flowers yellow.

ARALIACEÆ. (GINSENG FAMILY.)

Aralia spinosa. (PRICKLY ASH. HERCULES CLUB)
                          July—August. Flowers whitish.

CORNACEÆ. (DOGWOOD FAMILY.)

Cornus florida, L. (DOGWOOD.)   May. Involucre white.
    stricta, Lam.   April. Involucre white.
Nyssa multiflora, Wang. (SOUR GUM. TUPELO.) May. Flowers greenish.
    aquatica, L. (BLACK GUM.)   April—May. Flowers minute.
    uniflora, Walt. (COTTON GUM.) Deep swamps. Point Peter near
    R. R. bridge.   April. Flowers small, greenish.

CAPRIFOLIACEÆ. (HONEYSUCKLE FAMILY.)

Lonicera sempervirens, Ait. (WOODBINE.)
                  April—September. Flowers red.
Sambucus Canadensis, L. (ELDER.)   June—July. Flowers white.
Viburnum prunifolium, L. (BLACK HAW.)
                  April—May. Flowers small, white.
    nudum, L. (POSSUM HAW. SHAWNEE HAW.)   April—May.
    dentatum, L. (ARROWWOOD.)   March—May.

---

*Not given in Curtis' Catalogue.

RUBIACEÆ. (MADDER FAMILY.)

Galium pilosum, Ait.   June—September. Flowers purple.
    hispidulum, Michx.   May—September. Flowers greenish-white.
    trifidum, L. (SMALL BEDSTRAW.)   June—July. Flowers white.
    circœzans, Michx. (WILD LICORICE.)   July. Flowers purple.
Diodia Virginiana, L. (BUTTON WEED.) (*D. tetragona*.)
                    June—September. Flowers white or purplish.
    teres, Walter.   June—September. Flowers purplish.
Cephalanthus occidentalis, L. (BUTTON-BUSH. BOX.)
                              June. Flowers white.
Mitchella repens, L. (WILD RUNNING BOX.)
                    March—April. Flowers white.
Oldenlandia cœrulea, Gray. (BLUETS.)
                    February—March. Flowers pale blue.
    purpurea, Gray.   June—July. Flowers purple or nearly white.
    var. longifolia, Gray.
    glomerata, Michx.   July. Flowers greenish white.
Gelsemium sempervirens, Ait. (YELLOW JESSAMINE.)
                    March—April. Flowers yellow.
Spigelia Marilandica, L. (PINK ROOT.) Rare.   July. Flowers red.
Mitreola petiolata, T. & G. (MITREWORT.) Muddy banks.
                    June—September. Flowers small, white.
    sessilifolia, T. & G. Grassy swamps.   July—September.
Polypremum procumbens, L.   June—September. Flowers small, white.

COMPOSITÆ. (COMPOSITE FAMILY.)

Achillea millefolium, L. (MILFOIL. YARROW.)
                    June—September. Flowers white or rose.
Ambrosia artemisiæfolia, L. (RAGWEED. STICK-WEED. CARROT-WEED.)
                    August—September. Flowers unsightly.
Maruta COTULA, **D. C.** (*Anthemis cotula*.) (MAY WEED. FALSE CHAMOMILE.)
    Dr. McRee at F. Waddell's. Streets and roadsides.
                    June—September. Rays white.
Aster concolor, L. August—September. Rays purple, pappus rust colored.
    undulatus, L.   August—September. Flowers pale blue.
    dumosus, L.   September. Rays purplish white.
    flexuosus, Nutt. Salt marshes. August—October. Flowers purple.
    paludosus, Ait.   August—October. Rays violet blue.
    linifolius, L. Salt marshes.   October.
    Elliottii, T. & G.
    spectabilis, Ait.
    squarrosus, Walter.   September—November. Rays blue.
    patens, Ait.

simplex, Willd.　　　　　　　　　September. Rays pale blue.
miser, L.
Novi-Belgii, L.　　　　　　　　Rays pale blue or purplish.
carneus, Nees.
Novæ Angliæ, L.　　　　　　　　Rays violet purple.
longifolius, Lam.
Baccharis halimifolia, L. (GROUNDSEL.)
　　　　　　　　　September—October. Flowers white.
　glomeruliflora, Pers. (*B. Sessiliflora, Michx.*)
　　　　　　　　　November. Flowers white.
Baldwinia uniflora, Nutt. Brunswick county. Disk flowers, dark purple.
Bidens bipinnata, L. (SPANISH NEEDLES. BEGGAR-LICE.)
　　　　　　　　　August—September. Rays yellow.
　chrysanthemoides, Michx. (BUR-MARIGOLD.) September—October.
　frondosa, L.　　　　　　　　　　July—September.
Boltonia glastifolia, L'Her.　　　July—September. Rays white.
Pyrrhopappus Carolinianus, D. C. (*Barkhausia Caroliniana, Ell.*)
　　　　(FALSE DANDELION.)　　　April—July. Flowers yellow.
Borrichia frutescens, D. C. (SEA OX-EYE.) (*Buphthalmum frutescens. L.*)
　　Salt marshes.　　　　　　　June—October. Flowers yellow.
Chaptalia tomentosa, Vent. February—April. Flowers white or purplish.
Leucanthemum VULGARE, **Lam.** (WHITE DAISY. WHITE WEED.)
　　　(*Chrysanthemum leucanthemum, L.*)　Introduced. Found in
　　meadows. Common in vacant lots.
　　　　　　　　　May—July. Flowers white.
Bigelovia nudata, D. C. (*Chrysocoma nudata, Michx.*)
　　　　　　　　　August—September. Flowers yellow.
Chrysogonum Virginianum, L.　　February—March. Flowers yellow.
Chrysopsis graminifolia, Nutt. (SILVER GRASS. SCURVY GRASS.)
　　　　　　　　　August—September. Flowers yellow.
　Mariana, Nutt.　　　　　　　　　　September.
　gossypina, Nutt.　　　　　　September. Flowers yellow.
　var. dentata, Ell. Leaves larger than the above; lowest leaves semi-
　　ate toothed.
　trichophylla, Nutt.
*****Cirsium altissimum, Spreng.**　　August—September. Flowers purple.
　horridulum, Michx. (YELLOW THISTLE.)
　　　　Causeway between the ferries. April—May. Flowers often purple.
　repandum, Michx. Sand barrens.　　June—July. Flowers purple.
　†**Virginianum, Michx.**　　August—September. Flowers purple.
Coreopsis lanceolata, L.　　　　　May—June. Rays yellow.
　integrifolia, Poir.　　　　　September. Rays yellow.

---

*The generic name Cirsium is substituted for Cnicus, following Dr. Curtis'
Catalogue of Indig. Plants, p. 33, and Chapman's Botany, p. 247.

†Conyza Marilandica and C. bifrons, given in Plants around Wilmington, are
omitted from Curtis' Catalogue of Indig. Plants.

aurea, Ait. (*C. mitis, Michx.*)
August—October. Rays yellow, showy.

trichosperma, Michx. (TICK-SEED. SUNFLOWER.) September.

verticillata, L. (*C. tenuifolia. Ell.*) August. Disk yellow.

auriculata, L. June—September. Rays yellow, showy.

discoidea, T. & G. July—September.

angustifolia, Ait. September—October. Rays yellow.

*Eclipta erecta, L. September—October. Flowers white.

Elephantopus Carolinianus, Willd. (ELEPHANT'S FOOT.)
July—August. Flowers purple.

tomentosus, L. (*E. nudicaulis, Ell.*)
June.—August. Flowers pale purple.

Erigeron bellidifolium, Muhl. (ROBIN'S PLANTAIN.)
March—April. Rays blueish yellow.

Canadensis, L. (HORSE-WEED. HOG-WEED.¹)
May—September. Rays white.

Philadelphicum. L. (FLEABANE.) May. Rays purplish

vernum, T. & G. (*E. nudicaule, Michx.*) April. Rays white.

strigosum, Muhl. (DAISY FLEABANE.) June. Rays white or rose.

Eupatorium album, L. September.

aromaticum, L. (WILD HOREHOUND.) September. Flowers white.

coronopifolium, Willd. (DOG FENNEL.)
September—October. Flowers white.

fœniculaceum, Willd. (DOG FENNEL.)
September—October. Flowers white.

incarnatum, Walt. Dr. McRee. Rocky Point.
September. Flowers pale purple.

hyssopifolium, L. (*E. linearifolium, Walt.*) September.

perfoliatum, L. (WILD SAGE. BONESET. THOROUGHWORT.)
September. Flowers white.

serotinum, Michx. September.

rotundifolium, L. August.

teucrifolium, Willd. (*E. verbenæfolium, Michx.*)
September. Flowers white.

Gnaphalium polycephalum, Michx. (EVERLASTING.)
September—October. Flowers yellow.

purpureum, L. (CUDWEED.)
April—June. Purplish scales; corolla yellow.

Helenium autumnale, L. (SNEEZE-WEED.)
August—September. Flowers yellow.

quadridentatum, Labill. June—August. Rays yellow.

†tenuifolium, Nutt. G. McCarthy.

---

*The specific name procumbens is given by Dr. Curtis in his "Plants around Wilmington." The only two species described by Chapman (p. 224.) The plants are erect.

†Not given in Curtis' Catalogue.

Helianthus angustifolius, L. (Sunflower.)   October.   Flowers yellow.
   atrorubens, L. (*H. sparsifolius, Ell.*)
                           September—October.   Flowers yellow.
      heterophyllus, Nutt.
      giganteus, L.   September.   Rays yellow.
      annuus, L.* G. McCarthy.
Hieracium Gronovii, L.   September—October.   Flowers yellow.
   paniculatum, L.   August—September.   Flowers yellow.
Iva frutescens, L. (Marsh Elder.)   August—September.   Flowers whitish.
   imbricata, Walt. Sea coast.   August—September.   Flowers whitish.
Krigia Virginica, Willd.   March—May.   Flowers yellow.
   Caroliniana, Nutt.   February—March.   Flowers yellow.
Kuhnia Eupatorioides, L. (*K. critonia, Ell.*)
                           September.   Flowers yellowish white.
Lactuca elongata, Muhl. (Wild Lettuce.)
           July—September.   Flowers white, purple, blue or yellow.
      var. graminifolia, M. A. C.
†**Taraxacum Deus-Leonis, Desf.** (Dandelion.) (*Leontodon Taraxacum*)
                           May—August.   Flowers yellow.
Leptopoda puberula, Macbride.   April—May.   Disk yellow.
   ‡**fimbriatum, Gray.**   April—May.   Disk yellow.
§**Trilisa odoratissima, Cass.** (Vanilla-plant. Dog-tongue.)
                           July—August.   Flowers purple.
      paniculata, Willd.   September.   Flowers pale purple or white.
      spicata, Willd. (Button Snakeroot.)
                           August.   Flowers bright purple.
Cichorium Intybus, L. (Chiccory.)   Adventive—found in ballast.
                           July.   Flowers yellow.
Liatris squarrosa, Willd. (Blazing Star.)
                           July—August.   Flowers purple.
      tenuifolia, Nutt.   September.   Flowers purple.
      pauciflora, Pursh.   September.   Flowers purple.
      elegans, Willd.* G. McCarthy.   August.   Flowers purple.
      graminifolia, Pursh.   September.   Flowers purple.
Marshallia angustifolia, Pursh.
           July—August.   Flowers purplish with blue anthers.
      lanceolata, Pursh   April—June.   Flowers purplish.
Mikania scandens, Willd. (Climbing Hemp-weed.)
      (*M. pubescens, Muhl.*)   August—September.   Flowers whitish.

---

*Not given in Curtis' Catalogue.

†In "Plants around Wilmington," 1834, this plant is marked "rare." It is now very common in the steets and elsewhere, (1880.)

‡This is omitted in "Curtis' Catalogue of Indig. Plants."

§Formerly Liatris.

Polymnia Uvedalia, L. (BEAR'S-FOOT.)
July—August. Rays bright yellow.

Prenanthes alba, L. (*Nabalus albus, Hook.*) September. Flowers white.
virgata, Mich. (*Nabalus virgatus, D. C.*)
September. Flowers purplish.

altissima, L. (*Nabalus altissimus, Hook.*)
September. Flowers yellowish or greenish white.

Pterocaulon pycnostachyum, Ell. (BLACK-ROOT.)
June—July. Flowers white.

Melanthera hastata, Michx. August—September.

Rudbeckia hirta, L. (CONE FLOWER.) July—August. Rays yellow.

Senecio lobatus, Pers. March—April. Flowers yellow.
tomentosus, Michx. April—May. Flowers yellow.

Erechthites hieracifolia, Rafinesque. (FIREWEED.)
(*Senecio hieracifolius, L.*) July—September. Flowers greenish.

Silphium compositum, Michx. (ROSIN-WEED.)
See note in Gray's "Flora of North America."
July—September. Flowers yellow.

Solidago cæsia, L. September. Flowers yellow.
odora, Ait. (ANISE-SCENTED GOLDEN-ROD.)
October. Flowers yellow.

verna, M. A. C. Found on Duplin Road 3 to 5 miles from Wilmington, near Prigge's. See Map. May—June. Flowers yellow.

sempervirens, L. (*S. limonifolia, Pers.*)
September—October. Flowers yellow.

bicolor, L. September. Rays whitish.
tenuifolia, Pursh. October. Flowers yellow.
arguta, Ait. September. Flowers yellow.
var. juncea. September. Flowers yellow.
tortifolia, Ell. September. Flowers yellow.

*altissima, L. (*S. rugosa, Ell. S. Ulmifolia. S. Uspera.*)
September—October. Flowers yellow.

†angustifolia, Ell. October. Flowers yellow.
Elliottii, Tor. & Gray. September. Flowers yellow.
Boottii, Hook. September. Flowers yellow.
pilosa, Walt. (*S. pyramidata.*)
September—October. Flowers yellow.

puberula, Nutt. September. Flowers yellow.
var. pulverulenta. September. Flowers yellow.
petiolaris, Ait. September. Flowers yellow.
virgata, Michx. (*Syn. S. stricta.* Gray's Flora No. Am., Vol. I, Part II, p. 150.)
September. Flowers yellow.

---

*S. pilosa, S. recurvata, all synonymous with S. rugosa. See Gray's Flora No. Am., Vol. I, Part II, p. 153.

†Var. S. stricta. Gray's Flora of No. Am.' Vol. I, Part II, p. 150.

*Sonchus oleraceus, **L.** (Sow-thistle.) Introduced.
June—August. Flowers yellow.

Vernonia angustifolia, Michx.   June—August. Flowers purple.

Novæboracensis, Willd. (Iron-weed.)
July—September. Flowers purple.

Xanthium strumarium, **L.** (Cockle-bur.)   July—September.

spinosum, L. (Thorny Cockle-bur.)   August—September.

Conoclinium cœlestinum, D. C. (Mist-flower.
September. Flowers blueish purple.

Seriocarpus conyzoides, Nees. (White-topped Aster.)
(*Aster conyzoides.*)   August. Disk flowers yellow.

solidagineus, Nees. (*Aster solidaginoides.*)
August. Disk flowers yellow.

tortifolius, Nees. (Rattlesnake's Master.)
August. Disk flowers yellow.

Acanthospermum xanthioides, **D. C.**
Streets of Wilmington. Introduced since 1868. Nat. from So.
Am.   July—August. Flowers yellowish.

Pluchea bifrons, D. C. (*Conyza bifrons, Ell.*) Marsh Fleabane.)
September. Flowers purplish.

fœtida, D. C. (Stinking Fleabane.) [*C. Marilandica, Ell.*]
September. Flowers purple.

camphorata, D. C.   September. Flowers light purple.

Tanacetum vulgare, **L.** (Tansy.) Introduced. About settlements.
June—July. Flowers yellow.

Artemisia caudata, Michx. (Wild Wormwood.)   September.

Mulgedium acuminatum, D. C. (Blue Lettuce.) [*Sonchus Ell.*]
September. Flowers blue.

Spilanthes repens, Michx. Near Wilmington. August. Flowers yellow.

LOBELIACEÆ. (LOBELIA FAMILY.)

Lobelia cardinalis, L. (Cardinal Flower.) Rice fields and river swamps.
July—October. Flowers scarlet.

Nuttallii, R. & S. (*L. Kalmii, Ell.*)
August—September. Flowers pale blue.

puberula, Michx. (Blue Lobelia.)
August—September. Flowers bright blue.

syphilitica, L. (Great Lobelia.)
August—September. Flowers light blue.

Canbyi, Gray.

paludosa, Nutt.   May—August. Flowers white or pale blue.

glandulosa, Walt.   October. Flowers pale blue.

amœna, Michx.   September—October. Flowers bright blue.

---

*S. acuminatus and S. Carolinianus are not retained in Curtis' "Catalogue
of Indig. Plants." See p. 34.

CAMPANULACE.E. (CAMPANULA FAMILY.)

Campanula Americana, L. (BELL-FLOWER.)
August—September. Flowers blue.

ERICACEÆ. (HEATH FAMILY.)

Gaylussacia frondosa, Tor. & Gray. (BLUE HUCKLEBERRY.)
(*Vaccinium frondosa.*) April. Flowers white or reddish.
dumosa, T. & G. (DWARF HUCKLEBERRY.) [*V. dumosa.*]
April—May. Flowers white.
var. hirtella. April—May. Flowers white.
Vaccinium corymbosum, L. (SWAMP HUCKLEBERRY.) Common.
February—April. Flowers white.
tenellum, Ait. April. Flowers white.
arboreum, Mich. (SPARKLE BERRY.) May. Flowers white.
stamineum, L. (DEERBERRY. GOOSEBERRY.)
May—June. Flowers whitish.
myrsinites, Mich. Pine barrens. March—April. Flowers white.
crassifolium, And. (CREEPING HUCKLEBERRY.)
Common moist sandy savannahs and pine woods.
February—March. Flowers white or rose color.
Epigea repens, L. (TRAILING ARBUTUS. CROCUS.) .
Sand hills, half hidden under leaves of scrub oak.
February—March. Flowers white or rose color.
Gaultheria procumbens, L. (MOUNTAIN TEA. WINTER GREEN.)
June. Flowers white.
Leucothœ axillaris, Don. (*Andromeda axillaris, Lam.*)
February—March. Flowers white.
racemosa. (*A. racemosa, L.*) April—May. Flowers white.
Cassandra calyculata, Don. (*A. calyculata, L.*) April. Flowers white.
Andromeda nitida, Bartram. (FETTER BUSH.)
March—May. Flowers white, red or purple.
Mariana, L. (STAGGER BUSH.) April—May. Flowers white.
speciosa, Mich. May. Flowers white.
ligustrina, Muhl. (PEPPER BUSH.) (*A. paniculata.*)
May. Flowers white.
Oxydendrum arboreum, D. C. (SOUR WOOD. SORREL TREE.)
April—May. Flowers white.
Clethra alnifolia, L. (WHITE ALDER. SWEET PEPPER BUSH.)
May—October. Flowers white.
var. tomentosa. May—October. Flowers white.
Kalmia latifolia, L. (IVY.) Dr. McRee. Rocky Point.
May—June. Flowers white to deep rose color.
angustifolia, L. (WICKY.) May—October. Flowers rose red.
cuneata, Mich. Flowers white.

Azalea nudiflora, L. (Purple Honeysuckle.) Savannahs. Common.
April. Flowers white varying to rose color.
  viscosa, L. (Clammy Honeysuckle.)
          July—August. Flowers white.
Leiophyllum buxifolium, Ell. (Sand Myrtle.) Brunswick county.
          May. Flowers white.
Pyrola rotundifolia, L. (False Winter green.)
          June—July. Flowers white.
Chimaphila umbellata, Nutt. (Prince's Pine. Pipsissewa.)
          June. Flowers white.
  maculata, Pursh. (Spotted Winter green.) June. Flowers white.
Monotropa uniflora, S. (Eye-bright. Indian Pipe.) -
          August—September. Flowers white.

AQUIFOLIACEÆ. (HOLLY FAMILY.)

Ilex opaca, Ait. (Holly.)   April—May. Flowers white.
  verticillata, Gray.   April. Flowers white.
  Dahoon, Walt. var. myrtifolia, Chap. (Dahoon holly.)
          April—May. Flowers white.
  Cassine, L. [16] (Yaupon.) [*I. vomitoria.*]
      Found near Wilmington Dec., 1884, with yellow berries.
          April. Flowers white.
  glabra, Gray. (Gallberry. Inkberry.) May. Flowers white.
  coriacea, Chap. (Tall Gallberry.)   May. Flowers white.

STYRACACEÆ. (STORAX FAMILY.)

Styrax grandifolia, Ait. (Mock Orange.) April—May. Flowers white.
  Americana, Lam. (*S. glabrum, Ell.*)   May. Flowers white.
Halesia tetraptera, L. (Snow-drop Tree.)
          March—April. Flowers white.
Symplocos tinctoria, L'Her. (Yellow wood. Sweet leaf.)
  [*Hopea tinctoria.*]   March. Flowers yellow.

CYRILLACEÆ. (CYRILLA FAMILY.)

Cyrilla racemiflora, Walt. (Burn-wood Bark. He Huckleberry.)
          July. White flowers.

EBENACEÆ. (EBONY FAMILY.)

Diospyros Virginiana, L. (Persimmon.) May—June. Flowers greenish.

SAPOTACEÆ. (SAPODILLA FAMILY.)

Bumelia lycioides, Gært. (CAROLINA BUCKTHORN.)
Fruits September. Fruit a fleshy berry ovoid-black.
June—July. Flowers greenish.

PLANTAGINACEÆ. (PLANTAIN FAMILY.)

Plantago MAJOR, **L.** (PLANTAIN.) Introduced.
May—August. Flowers white.
LANCEOLATA, L. (NARROW LEAVED PLANTAIN.)
May—August. Flowers white.
Virginica, L. April—June.
sparsiflora, Mich. (*P. interrupta, Lam.*) June—September.

PLUMBAGINACEÆ. (LEADWORT FAMILY.)

Statice Caroliniana, Walt. (MARSH ROSEMARY.) [*S. limonium.*]
August—September. Flowers blue.

PRIMULACEÆ. (PRIMROSE FAMILY.)

Lysimachia stricta, Ait. (LOOSE STRIFE.) Walker's causeway.
July. Flowers yellow.
ciliata, L. Dr. McRee. Rocky Point,
July—August. Flowers yellow.
Anagallis ARVENSIS, **L.** Introduced. Rare. July. Flowers red.
Samolus floribundus, Kunth. (BROOK-WEED.) [*S. valerandi.*]
May—July.
valerandi var. Americana,* W. M. C. Fort Fisher.

LENTIBULACEÆ. (BLADDERWORT FAMILY.)

Utricularia inflata, Walt. (BLADDERWORT.)
April—May. Flowers yellow.
striata, LeConte. September. Flowers yellow.
fibrosa, Walt. [*M. longirootris, LeConte.*]
May—June. Flowers yellow.
cornuta, Mich. [*M. personata, LeConte.*]
July—September. Flowers yellow.
permeata,* LeConte. Ger. McCarthy, near McIlhenny's mill pond.
August. Flowers yellow.
subulata, L. [*M. Setacea, Michx.*] February—May. Flowers yellow.
personata,* Vahl. Ger. McCarthy.
purpurea, Walt. June. Flowers purple.

---

*Not given in Curtis' Catalogue.

Pinguicula lutea, Walt. (BUTTERWORT.)

February—April. Flowers yellow.

    elatior, Michx.   March—April. Flowers purple to white.

    pumila,* Michx. McIlhenny's pond.

March—April. Flowers purple to white.

BIGNONIACEÆ. (BIGNONIA FAMILY.)

Bignonia Capreolata, L. (CROSS-VINE.)   April. Flowers red.

Tecoma radicans, Jussieux. (TRUMPET-FLOWER.) [*B. radicans.*]

May—June. Flowers scarlet.

Catalpa BIGNONIOIDES, **Walt.** (CATALPA.) Introduced.

May. Flowers white.

Martynia PROBOSCIDEA, **Glox.** (MARTINO. UNICORN PLANT.)

June—August. Flowers white.

OROBANCHACEÆ. (BROOM-ROPE FAMILY.)

Epiphegus Virginiana, Bartram. (BEECH-DROPS.) Rev. Mr. Hunt.

August. Flowers purplish.

Conopholis Americana, Wall'r. (SQUAW-ROOT.)

Dr. McRee. Rocky Point. [*Orobranche Americana.*]

April. Flowers yellowish.

SCROPHULARIACEÆ. (FIGWORT FAMILY.)

Verbascum THAPSUS, **L.** (MULLEIN.) Introduced.

May—August. Flowers yellow.

    BLATTARIA, L. (MOTH MULLEIN.)   May—August. Flowers yellow.

Scrophularia nodosa, L. (FIG WORT.) [*S. Marilandica, L.*]

September. Greenish purple.

Chelone glabra, L. (SNAKE-MOUTH.) July. Flowers white or rose color.

    †**var. purpurea.**   July. Flowers purple.

Pentstemon pubescens, Soland. (BEARD-TONGUE.)

June—July. Flowers purple.

    var. lævigatus.   June—July. Flowers purple.

Linaria Canadensis, Spreng. (TOAD FLAX.) [*Antirrhinum Canadense, L.*]

April.—May. Flowers blue and white.

    SPURIA, L. Introduced.

Mimulus ringens, L. (MONKEY-FLOWER.)   August. Flowers showy.

---

*Not given in Curtis' Catalogue.

†Not repeated in Curtis' "Catalogue Indig. Plants."

Herpestis nigrescens, Benth. [*Gratiola acuminata, Walt.*]
August—September. Flowers striped with blue.

Monniera, H. B. & K. (*H. cuneifolia.*) Fort Fisher.
June—September. Flowers white or pale blue.

amplexicaulis, Pursh. July—September. Flowers blue.

Gratiola Virginiana, L. (HEDGE HYSSOP.) April—May. Flowers white.

spærocarpa, Ell. (*G. Caroliana.*) March—May. Flowers white.

pilosa, Mich. June—August. Flowers white.

Ilysanthes gratioloides, Benth. (FALSE PIMPERNEL.)
(*Lindernia attenuata. Gratiola tetragona.*)
May—September. Flowers small, purplish.

Micranthemum orbiculatum, Mich. June—October. Flowers white.

Veronica ARVENSIS, L. (CORN SPEEDWELL.)
May—June. Flowers pale blue.

peregrina, L. (PURSLANE SPEEDWELL.)
April—June. Flowers white.

serpyllifolia, L. (PAUL'S BOTANY.) May—September. Flowers blue.

Seymeria tenuifolia, Pursh. August—September. Flowers yellow.

Dasystoma pubescens, Benth. (FALSE FOXGLOVE.) (*Gerardia flava, L.*)
July—September. Flowers yellow.

pectinata, Benth. (*G. pectinata.*)
August—September. Flowers yellow.

Gerardia linifolia, Nutt. (FLAX-LEAVED GERARDIA.)
September. Flowers purple.

aphylla, Nutt. September. Flowers purple.

divaricata, Chap. September. Flowers purple.

purpurea, L. (PURPLE GERARDIA.) September. Flowers purple.

var. fasciculata. September. Flowers purple.

maritima, Raf. Sea beach. September. Flowers purple.

tenuifolia, Vahl. September. Flowers purple.

setacea, Ell. (*G. Plunkettii, Ell.*) September. Flowers purple.

Schwalbea Americana, L. (CHAFF-SEED.)
May—June. Flowers yellow and purple.

ACANTHACE.E. (ACANTHUS FAMILY.)

Dipteracanthus strepens, Nees. (*Ruellia strepens, L.*)
June—September. Flowers blue or purple.

Dianthera ovata, Walt. (WATER WILLOW.) (*Justicia humilis.*)
July—August. Flowers pale blue

VERBENACEÆ. (VERVAIN FAMILY.)

Verbena Caroliniana, Michx.   August. Flowers pink.

OFFICINALIS, L. (VERVAIN.) . Introduced. (*V. spuria, L.*)
                                         July—August. Flowers purple.
urticifolia, L. (WHITE VERVAIN.) Rocky Point. Dr. McRee.
                            August—October. Flowers white or pale blue.
Lippia nodiflora, Michx. (FOG-FRUIT.) (*Zapania nodiflora, Lam.*)
                            May—September. Flowers white or purple.
Callicarpa Americana, L (BERMUDA MULBERRY.)
                                         June—July. Flowers blue.

LABIATÆ. (MINT FAMILY.)

Hyptis radiata, Willd.   July—September. Flowers white.
Mentha VIRIDIS, L. (SPEARMINT.) Introduced. (*M. tenuis, Michx.*)
                            July—September. Flowers pale blue.
PIPERITA, L. (PEPPERMINT.) Introduced.
                            July—September. Flowers white or blue.
ROTUNDIFOLIA, L. (ROUND-LEAVED MINT.) Rare.
            Causeway Kidder's rice field. Wharves. July. Flowers white.
Lycopus Virginicus, L. (BUGLE-WEED.)   September. Flowers white.
    sinuatus, Ell. (*L. exaltatus, Ell*)   July.
Pycnanthemum aristatum, Michx. August—September. Flowers white.
    incanum, Michx. (MOUNTAIN MINT.)
                            August—September. Flowers white.
        linifolium, Pursh. Rocky Point. Dr. McRee.
                                         August—September.
Collinsonia punctata, Ell.   September. Flowers yellowish.
Calamintha NEPETA, **Link.** Common about the streets. Introduced.
                            July—September. Flowers purple.
Salvia lyrata, L.   April—May. Flowers blue.
Monarda punctata, L (RIGNUM.)
                August—October. Flowers yellowish with purple bracts.
Nepeta CATARIA, **L.** (CATNIP.) Introduced
                            July—September. Flowers white.
Scutellaria integrifolia, L.   May—July. Flowers blue.
    pilosa Michx.   July—August. Flowers pale blue.
    nervosa,* Pursh. Ger. McCarthy.   July.
    serrata, Andr.   July. Flowers blue.
Macbridea pulchra, Ell. Rare. Point Peter causeway.
                            August—September. Flowers purple.

---

*Not given in Curtis' Catalogue.

Physostegia Virginiana, Benth. (DRAGON-HEAD.)
    (*Dracocephalum obovatum, Ell.*)
             June —August. Flowers purplish.

Lamium AMPLEXICAULE, **L.** (DEAD-NETTLE HEN-BIT.) Introduced.
    Common in gardens, fields and highways. May. Flowers purple.

Marrubium VULGARE, **L.** (HOREHOUND.) Introduced.
            ● July—September. Flowers white.

Leonurus CARDIACA, **L.** (MOTHER-WORT.) Introduced.

Stachys hyssopifolia, Mich. (HEDGE-NETTLE.)
             June—August. Flowers violet.

Trichostema dichotomum, L. (BLUE-CURLS.)
             August—September. Flowers blue.

Teucrium Canadense, L. (WOOD-SAGE.)
             July—September. Flowers purplish.

BORRAGINACEÆ. (BORAGE FAMILY.)

Heliotropium Curassavicum, L. (HELIOTROPE.) Near the coast.
             June August. Flowers white.
    EUROPEUM, L.* G. McCarthy.   July.

Onosmodium Virginianum, D. C. (*O. hispidum, Michx.*)
             May—June. Flowers greenish.

Heliophytum INDICUM, **D. C.** (*Heliotropum Indicum.*)
    (INDIAN HELIOTROPE.) Introduced.
             June—October. Flowers blue.

Lithospermum ARVENSE, **L.** (CORN GROMWELL.) Belvidere.
             May—April. Flowers yellowish white.

Echium VULGARE, **L.** (BLUEWEED. VIPER'S BUGLOSS.)
    Common in streets, gardens and waste places.
             June—September. Flowers blue or purple.

HYDROPHYLLACEÆ. (WATER-LEAF FAMILY.)

Phacelia parviflora, Pursh.   April—May. Flowers pale blue or white.

HYDROLEACEÆ. (HYDROLEA FAMILY.)

Hydrolea quadrivalvis, Walt.   July—August. Flowers blue.

POLEMONIACEÆ. (POLEMONIUM FAMILY.)

†**Phlox paniculata, L.** (PHLOX.) Near Eden's Mills. Common in streets.
             June—July. Flowers purple or white.
    ‡**subulata, L.** (WILD PINK.) April—May. Flowers purple or white.

---

*Not given in Curtis' Catalogue.

†Phlox paniculata is put down in Curtis' "Catalogue of Indig. Plants " as found in Lincoln county and westward.

‡P. setacea is the same as P. subulata.

13

Pyxidanthera barbulata, Mich. (FLOWERING MOSS.)
 Common in savannahs. [*Diapensia barbulata, Ell.*]
 March—April. Flowers white.

CONVOLVULACEÆ. (CONVOLVULUS FAMILY.)

•

Quamoclit COCCINEA, **Mœnch.** (CYPRESS VINE.)
 Common in cultivated grounds.
 July—August. Flowers sometimes yellowish scarlet.
Pharbitis NIL, **Chois.** (MORNING GLORY.)
 July—September. Flowers purple.
Ipomœa commutata, R. & S. (*I. trichocarpa, Ell.*)
 August—October. Flowers purple.
 pandurata, Meyer. (WILD POTATO.)
 August—October. Flowers white.
 sagittifolia, Bot. Reg. (*Convolvulus sagittifolius, Michx.*)
 Smithville. July—September. Flowers purple.
 lacunosa, L. August—October. Flowers white.
Calystegia PARADOXA, **Pursh.** Flowers white.
Stylisma humistrata, Chap. (*Convolvulus tenellus, Ell.*)
 July—September. Flowers white.
 aquatica, Chap. July—September. Flowers purple.
 Pickeringii, Gray. July—September. Flowers white.
Dichondra repens, Forster. var. Caroliniensis, Chois.
 March—October. Flowers greenish white.
Cuscuta arvensis, Beyr. (LOVE VINE DODDER.)
 June—July. Flowers yellowish.
 Gronovii, Willd. (*C. Americana, Pursh, D. C.*)
 August—September. Flowers white.
 compacta, Juss. July—October. Flowers whitish.

SOLANACEÆ. (NIGHTSHADE FAMILY.)
•

Solanum nigrum, L. (NIGHTSHADE.) Common near settlements.
 July—September. Flowers white.
 Caroliuense, L. (HORSE-NETTLE.)
 June—September. Flowers blue or white.
Physalis viscosa, L. (GROUND CHERRY.) July—October. Flowers yellow.
 lanceolata, Michx. On the coast.
 July—October. Flowers yellow in the throat.
 angulata, L. Waste grounds. July—October. Flowers yellow.
 pubescens, L. Waste grounds.
 July—October. Flowers bright yellow.
Datura STRAMONIUM. (JAMESTOWN-WEED. JIMSON-WEED.)
 June—October. Flowers white.
 var. Tatula. Flowers purplish.
 QUERCIFOLIA, H. B. K. Ballast. August. Flowers white.
 METEL, L. Lower District. Dr. McRee. Causeway.
 July. Flowers white.

WRIGHTII. Escaped from cultivation. Smithville.
> July. Flowers white.

Petunia ———. Found in dumping grounds. Escaped from cultivation.
> July. Flowers white.

GENTIANACEÆ. (GENTIAN FAMILY.)

Sabbatia angularis, Pursh. (CENTAURY.) Savannas.
> July—August. Flowers purple.

brachiata, Ell.
> July—August. Flowers purple.

calycosa, Pursh.
> July—August. Flowers white.

paniculata, Pursh.
> August. Floweis reddish, margins of petals and throat of corolla sometimes distinctive pink.

gentianoides, Ell.
> July—August. Flowers purple.

stellaris, Pursh. Salt marshes. August—September. Flowers white.

Gentiana Elliottii, Chap. (SAMPSON SNAKE-ROOT.)
> October. Flowers bright blue.

Saponaria, L. (SAMPSON SNAKE-ROOT.)
> September—October. Flowers light blue.

angustifolia, Michx. (NARROW-LEAVED GENTIAN.)
> November—December. Flowers green without, white within.

Bartonia tenella, Muhl. September—October. Flowers greenish white.

verna, Muhl. (*Centaurella verna, Michx.*) Moist sandy places.
> February—April.

Obolaria Virginica, L. Rocky Point. Dr. McRee.
> March—April. Flowers purplish.

Limnanthemum lacunosum, Griseb. (FLOATING HEART.)
> June—July. Flowers white.

trachyspermum, Gray.
> April—June. Flowers white.

APOCYNACEÆ. (DOGBANE FAMILY.)

Apocynum cannabinum, L. (INDIAN HEMP.)
> Dr. McRee. Rocky Point, Masonboro Sound.
> July—August. Flowers greenish white.

Forsteronia difformis, A. D. C. (*Echites difformis, Walt.*)
> May—August. Flowers yellow.

Amsonia Tabernæmontana, Walt. (*A. latifolia, Michx.*)
> May—June. Flowers pale blue.

ASCLEPIADACEÆ. (MILKWEED FAMILY.)

Asclepias amplexicaulis, Michx. (RABBIT'S MILK.)
> April—May. Flowers ash color.

obtusifolia, Michx.
> June—July. Flowers greenish purple.

paupercula, Michx.
> June—July. Flowers deep red.

quadrifolia, Jacq.
> June—August. Flowers pale pink.

tuberosa, L. (BUTTERFLY-WEED. PLEURISY-ROOT.)
> June—July. Flowers orange.

variegata, L.   May—June. Flowers white.
verticillata, L.   July—September. Flowers greenish.
incarnata, L. (SWAMP SILK-WEED.)
                                   June—July. Flowers reddish purple.
Acerates longifolia, Ell.   July. Flowers pale purple.
Podostigma pubescens, Ell.   .   June—October. Flowers orange.
Seutera maritima, Decaisne. (*Vincetoxicum palustre.*) Salt marshes.
                                   July—August. Flowers greenish.
Gonolobus hirsutus, Mich. (RUNNING MILK-WEED.)
                                        September. Flowers white.
    macrophyllus, Mich.   July—August. Flowers purplish.

OLEACEÆ. (OLIVE FAMILY.)

Olea Americana, L. (DEVIL-WOOD. AMERICAN OLIVE.) Near the coast.
                                   March—April. Flowers white.
Ligustrum VULGARE, L. (PRIVET.) Introduced. Cultivated.
                                   May—June. Flowers white.
Chionanthus Virginica, L. (FRINGE-TREE. OLD MAN'S BEARD.)
                                   April—May. Flowers white.
Fraxinus platycarpa, Mich. (WATER ASH.)
                                   March—April. Flowers white.

ARISTOLOCHIACEÆ. (BIRTHWORT FAMILY.)

Aristolochia Serpentaria, L. (VIRGINIA SNAKE-ROOT.)
                                   June.—August. Flowers dull purple.

PHYTOLACCACEÆ. (POKEWEED FAMILY.)

Phytolacca decandra, L. (POKE-WEED.)
                                   July—September. Flowers white.

CHENOPODIACEÆ. (GOOSEFOOT FAMILY.)

Chenopodium ALBUM, L. (LAMB'S QUARTERS.)
                                   July—September. Flowers greenish.
    BOTRYS, L..* Ger. McCarthy.   August. Flowers greenish.
    ANTHELMINTICUM, L. (WORM-SEED. JERUSALEM OAK.)
        Introduced.   August. Flowers greenish.
    AMBROSIOIDES, L..   August. Flowers greenish.
    MURALE, L.   Flowers greenish.

---

*Not given in Curtis' Catalogue.

Atriplex hastata, L. (ORACHE.) Sea shore.   June—September.
Obione arenaria, Moq. (SAND ORACHE.) Sea beach. July—September.
Chenopodina maratima, Moq. (SEA GOOSEFOOT.) Salt marshes.
         September. Flowers minute.
Salicornia herbacea, L. (SAMPHIRE.) Salt marshes.   August.
 ambigua, Michx. Salt marshes.   August.
Salsola Kali, L. (SALTWORT.) Sandy sea shore.
         August. Flowers rose color.

AMARANTACEÆ. (AMARANTH FAMILY.)

Amarantus HYBRIDUS, **L.** (GREEN AMARANTH. CARELESS.) Introduced.
    August—September. Flowers greenish or purplish.
 SPINOSUS, L. (THORNY AMARANTH.)
     July—October. Flowers greenish or purplish.
 CHLORASTACHYS, WILL. Cultivated ground.
     August—September. Flowers greenish.
Euxolus pumilus, Raf. (DWARF AMARANTH.) Sandy sea shore.
     August—September. Flowers greenish.
Acnida cannabina, L. (WATER HEMP.)
Telanthera polygonoides, Moq.
Alternanthera Achyrantha, R. Br.   June—October.

POLYGONACEÆ. (BUCKWHEAT FAMILY.)

Polygonum ORIENTALE, **L.** (PRINCE'S FEATHER.) Introduced.
   June—September. Flowers large, bright rose color.
 Pennsylvanicum, L. July—September. Flowers rose color.
 PERSICARIA, L. (LADY'S THUMB.) July. Flowers rose color.
 acre, Kth. (SMARTWEED.) [*P. punctatum, Ell.*] July—September.
 hydropiperoides, Mich. (WATER-PEPPER.) [*P. mite, Pers.*]
    July—September. Flowers pale rose color.
 hirsutum, Walt.   July—September. Flowers white.
 aviculare, L. (KNOT-GRASS.) Flowers greenish white.
 setaceum, Bald.* July—September. Flowers white.
 †**littorale.** [*P. Maritimum, L.*] Sea beach. Flowers reddish white.
 incarnatum, Ell.* G. McCarthy.
    July—October. Flowers small, flesh color.
 Virginianum, L.   August—September. Flowers greenish.
 arifolium, L. (SCRATCH-GRASS.) June—October. Flowers white.
 sagittatum, L. (TEAR-THUMB.) June—October. Flowers white.

---

*Not given in Curtis' Catalogue.

†See Chapman's "Flora of the Southern U. S," p. 390.

Polygonella parvifolia, Mich.
        August—September. Flowers white, yellowish or rose color.
    articulata, Meisn.   August. Flowers bright rose color.
Rumex CRISPUS, **L.** (SOUR DOCK.)   June—July.
    verticillatus, L. (SWAMP DOCK.) [*R. Brittanicus, Ell.*]
                                    May—June.
    ACETOSELLA, L. (SORREL.)   June—July. Flowers small.
    OBTUSIFOLIUS, L. (BITTER DOCK.)   July—August.
    Engelmanii, Ledeb.
    maritimus, L. (GOLDEN DOCK.) Sea shore. Rare.
                August—September. Spikes yellowish.

LAURACEÆ (LAUREL FAMILY.)

Persea Carolinensis, Nees. (RED BAY.) [*Laurus Carolinensis, L.*]
    var. palustris, M. A. C.   July.
Sassafras officinale, Nees. (SASSAFRAS.) [*Laurus sassafras.*]
        March. Flowers sometimes white.
Benzoin odoriferum, Nees. (SPICE-BUSH. FEVER-BUSH.)
        February—March. Flowers yellow.
    melissæfolium, Nees. Dr. McRee. [*L. melissæfolia. Walt.*]
        February—March. Flowers yellow.
Tetranthera geniculata, Nees. (POND BUSH.)
        February—March. Flowers yellow.

LORANTHACEÆ. (MISTLETOE FAMILY.)

Phoradendron flavescens, Nutt. (MISTLETOE.)
    [*Viscum flavescens, Pursh.*]   April—May.

SAURURACEÆ. (LIZARD'S-TAIL FAMILY.)

Saururus cernuus, L. (SWAMP LILY. LIZARD'S-TAIL.)
        May—August. Flowers white.

CERATOPHYLLACEÆ. (HORNWORT FAMILY.)

Ceratophyllum demersum, L.   September—October.

CALLITRICHACEÆ. (WATER STARWORT FAMILY.)

Callitriche verna, L. (WATER STARWORT.)   March—April.

EUPHORBIACEÆ. (SPURGE FAMILY.

Euphorbia corollata, L. (Flowering Spurge.)   July—September.
    var. angustifolia, M. A. C.   July—September.
    Cyparissias, L.* Advent from Europe. Ger. McCarthy.
    obtusata, Pursh. (Warted Spurge.)   May.
    serpyllifolia, Pursh.* Ger. McCarthy.
    cyatobphora, Jacq. Streets.   May—October.
    Ipecacuanhæ, L. (Wild Ipecac.) Sand hills.
        May—June. Flowers not conspicuous, greenish yellow.
    maculata, L. (Spotted Spurge.) Common in waste grounds.
                              June—October.

    marginata, Pursh. (Variegated Spurge.)
    polygonifolia, L. (Shore Spurge.) Sea shore.   July—October.
    forbuserpens, H. B. K.* Ger. McCarthy.
    Curtisii, Englemann. Sea coast.   August.
    cordifolia, Ell.   July—September.
Stillingia sylvatica, L. (Queen's Delight.)   April—September.
    ligustrina, Mich. Dr. McRee.   May—August.
†Acalypha Virginica, L. (Three-seeded Mercury.) July—September.
    gracilens, Gray.   July—September.
Tragia urens, L. (Nettle.)   May—August. Flowers minute greenish.
Croton maritimum, Walt. Sea coast.   July—October.
Cnidosculus stimulosus, Gray. (Tread softly. Nettle Potatoe.)
    [*Iatropha stimulosa, Mich.*]   April—September. Flowers white.
Ricinus communis, L. (Castor-oil Plant.) Introduced.
                    June—October. Flowers yellow.

URTICACEÆ. (NETTLE FAMILY.)

Urtica urens, L. (Stinging Nettle.)
                December—February. Flowers greenish.
    capitata, Willd.   July—August. Flowers greenish.
Pilea pumila, Gray. (Clear-weed.)   July—September.
Bœhmeria cylindrica, Willd. (False Nettle.)   July—September.

MORACEÆ. (MULBERRY FAMILY.)

Morus rubra, L. (Mulberry.)   March.
    alba, L. (White Mulberry.) Introduced.
Broussonetia papyrifera, Vent. (Otaheite Mulberry.)

---

*Not given in Curtis' Catalogue.

†Acalypha Caroliniana, Walt., not repeated in Curtis' "Cat. Indig. Plants."

ULMACEÆ. (ELM FAMILY.)

Ulmus Americana, L. (ELM.)
February—March. Flowers greenish or purplish.
fulva, Mich. (SLIPPERY ELM.)
February—March. Flowers greenish or purplish.
alata, Mich. (SMALL-LEAVED ELM. CORK-WINGED ELM.)
Planera aquatica, Gmel. (PLANER-TREE.) South of Cape Fear river.
February—March.
Celtis occidentalis, L. (HACKBERRY.) March. Flowers greenish.
var. pumila, M. A. C. (DWARF HACKBERRY.)
March. Flowers greenish.

PLATANACEÆ. (PLANE-TREE FAMILY.)

Platanus occidentalis, L. (SYCAMORE.)   March—April.

JUGLANDACEÆ. (WALNUT FAMILY.)

Juglans nigra, L. (BLACK WALNUT.)   March—April.
Carya tomentosa, Nutt. (WHITE HICKORY.)   March—April.
amara, Nutt. (BITTER-NUT HICKORY.)   March—April.

CUPULIFERÆ. (OAK FAMILY.)

Quercus alba, L. (WHITE OAK.)
aquatica, Cates. (WATER OAK.)
Catesbæi, Mich. (SCRUB OAK.)
cinerea, var. pumila, Michx. (RUNNING OAK. DWARF OAK.)
lyrata, Walt. (OVERCUP OAK.)
coccinea, Wang. (SCARLET OAK.)
falcata, Mich. (SPANISH OAK.)
nigra, L. (BLACK JACK.)
obtusiloba, Mich. (POST OAK.)
Phellos, L. (WILLOW OAK.)
Prinus, L. (SWAMP CHESTNUT OAK.)
prinoides, Willd. (CHINQUAPIN OAK.)
virens, Ait. (LIVE OAK.)
Castanea pumila, L. (CHINQUAPIN.)   April—May.
Fagus ferruginea, Ait. (BEECH.)   April.
Carpinus Americana, Michx. (HORNBEAM.)   March.

MYRICACEÆ. (WAX MYRTLE FAMILY.)

Myrica cerifera, L. (WAX MYRTLE. BAYBERRY.) Swamps. March—April.
    var. pumila. Barrens. Sand hills.

BETULACEÆ. (BIRCH FAMILY.)

Betula nigra, L. (RED BIRCH.) [*B. rubra, Michx.*]   March.
Alnus serrulata, Ait. (ALDER.)   January—March.

SALICACEÆ. (WILLOW FAMILY.)

Salix nigra, Marshall. (BLACK OR SWAMP WILLOW.)
    BABYLONICA, TOURN. (WEEPING WILLOW.) About dwellings.
Populus angulata, Ait. (CAROLINA POPLAR.)   March—April.
    heterophylla, L. (COTTON TREE.)   March—April.
    DILATATA, AIT. (LOMBARDY POPLAR.) Introduced.

CONIFERÆ. (PINE FAMILY.)

Pinus mitis, Michx. (YELLOW OR SHORT-LEAVED PINE.)
    Tæda, L. (OLD FIELD, LOBLOLLY OR SLASH PINE.)
    australis, Michx. (LONG-LEAF PINE.)
        [*P. palustris, L.*, the prior but inappropriate name. Chapman.]
    Serotina, Michx. (POND-PINE.)
Cupressus thyoides, L. (WHITE CEDAR. JUNIPER.)   April.
Taxodium distichum, Rich. (CYPRESS.) [*C. distichum.*] February—March.
    var. imbricaria. Cypress pond marked on Map.
Juniperus Virginiana, L. (RED CEDAR.)   March.

14

ENDOGENS.

PALMACEÆ. (PALMS.)

Sabal Palmetto, R. & S. (Palmetto.) Cape Fear.   June.
    Adansonii, Guerns. (Dwarf Palmetto.)
      Eagle's Island causeway.   June—July.

ARACEÆ. (ARUM FAMILY.)

Arisæma triphyllum, Torr. (Indian Turnip.)   March.
Xanthosoma sagittifolium, Schott. (Arrow-leaved Spoon Flower.)
    "Branches" and spongy ground. Eastern side of Wilmington
    and Brunswick. Point Peter causeways. Fruit ripens Sept.
    15 to 30. Beautiful crimson. [*Caladium glaucum*, (?) *Ell.*]
                May—June. Flowers white.
Orontium aquaticum, L. (Golden Club. Water Dock.)
    Tuckahoe—a purely local name without any good foundation for
    it; probably a corruption of Tuckaw. See Kalm's Travels.
                       March—April.
Acorus Calamus, L. (Calamus.)   April.

LEMNACEÆ. (DUCKWEED FAMILY.)

Lemna minor, L. (Duckweed.)
    polyrhiza, L.

TYPHACEÆ. (CAT-TAIL FAMILY.)

Typha latifolia, L. (Cat-tail.)   July—August.
Sparganium ramosum, Hudson. (Bur-reed.)   July.

NAIADACEÆ. (POND WEED FAMILY.)

Zostera marina, L. (Sea-wrack. Eel-grass.) Salt water.
                    August—September.
Ruppia maritima, L. (Ditch-grass.)   May—August.
Potamogeton pectinatus, L. (Pondweed.)   June—August.
    lucens, L.   August.
    hybridus, Michx.   June—August.

heterophyllus, Schreb.   July.
pan iflorus, Pursh.   July—August.
perfoliatus, L.   July—September.
fluitans, Roth.   June—August.
Naias flexilis, Rotsk.

ALISMACEÆ. (WATER-PLANTAIN FAMILY.)

Alisma plantago, L. (WATER-PLAINTAIN.)   July—August.
Triglochin triandrum, Michx. (ARROW-GRASS.)   August—September.
Echinodorus radicans, Engelmann. (*Alisma radicans, Nutt*)
                   July—September.
*Sagittaria variabilis, Eng. (ARROW-LEAF.)   July—September.
    falcata, Pursh.   June—September.
    pusilla, Nutt.
    heterophylla, Pursh.
    natans, Michx.   June—September.
    simplex, Nutt.   May—October.
    graminea, Michx.† Ger. McCarthy.

HYROCHARIDACEÆ. (FROG'S-BIT FAMILY.)

Limnobium Spongia, Rich. (FROG'S-BIT.)   July—August.

BURMANNIACEÆ. (BURMANNIA FAMILY.)

Burmannia biflora, L.   October. Flowers pinkish.
    capitata, Chap. (*Tripterella capitata, Michx.*)
                   October. Flowers white.

ORCHIDACEÆ. (ORCHIS FAMILY.)

Microstylis orphioglossoides, Nutt. (ADDER'S-MOUTH.)
             July—August. Flowers greenish.
‡Bletia aphylla, Nutt. Dr. McRee. Rocky Point.
       July—August. Flowers brownish, striped with purple.
Pouthieva glandulosa, R. Br. (*Ophrys pubera, Michx.*)
             September—October. Flowers greenish.
Corallorhiza odontorhiza, Nutt. (CORAL-ROOT.)
    [*Cymbidium corallorhizoa.*]   February—March.

---

*S. variabilis variously named S. sagittifolia, hastata, pubescens, etc.

†Not given in Curtis' Catalogue.

‡A specimen found by Mr. Charlie Bradley in his boathouse at Greenville
Sound, on Hewlett's Creek, 1886.

Habenaria repens, Nutt.   August—September. Flowers small, greenish.
Listera australis, Lindley. (TWAYBLADE.) July. Flowers small, greenish.
Pogonia divaricata, R. Br.   May. Flowers flesh-colored.
   ophioglossoides, Nutt.   April—May. Flowers pale rose color.
Tipularia discolor, Nutt. (CRANE-FLY ORCHIS.)
                                    August. Flowers greenish.
Platanthera flava, Gray. (YELLOW ORCHIS.)
                  July—August. Flowers brownish green.
   ciliaris, Lindley. (YELLOW FRINGED ORCHIS.)
                        Flowers large bright yellow.
   blephariglottis, Lindley. (WHITE FRINGED ORCHIS.)
                            August. Flowers white.
Spiranthes cernua, Rich. (LADY'S TRESSES.)
                  October. Flowers yellowish white.
   odorata, Nutt.   October. Flowers yellowish white.
   tortilis, Willd.   May. Flowers white.
   gracilis, Bigelow.   April—May. Flowers minute.
Epidendrum* conopseum. Major Young, Pender county, 1881.
                August. Flowers green, tinged with purple.
Calopogon pulchellus, R. Br. (BEARDED PINK.)
                    June. Flowers bright purple.
   parviflorus, Lindley.   March—April. Flowers bright purple.
     var. albus.

AMARYLLIDACEÆ. (AMARYLLIS FAMILY.)

Amaryllis Atamasco, L. (ATAMASCO LILY. STAGGER-GRASS.)
         March—April. Flowers white, tinged with red.
Pancratium rotatum, Ker. (*P. Mexicanum.*) April—May. Flowers white.
Hypoxis erecta, L. (YELLOW STAR-GRASS.)
                March—April. Flowers yellow.

HÆMODORACEÆ. (BLOODWORT FAMILY.)

Lachnanthes tinctoria, Ell. (RED-ROOT.)
               July—September. Flowers yellow.
Aletris farinosa, L. (STAR-GRASS. COLIC-ROOT.)
           May—June. Flowers white or yellow.
   aurea, Walt.   May—June. Flowers white or yellow.

BROMELIACEÆ. (PINE-APPLE FAMILY.)

Tillandsia usneoides, L. (LONG MOSS.) June—September. Petals green.

---

*Not given in Curtis' Catalogue.

IRIDACEÆ. (IRIS FAMILY.)

Iris versicolor, L. (BLUE FLAG.) April—May. Flowers pale blue.
 tripetala, Walt. June—July. Flowers blue.
 Virginica, L. June. Flowers blue or white.
 verna, L. (DWARF IRIS.) April. Flowers pale blue.
Sisyrinchium Bermudiana, L. (BLUE-EYED GRASS. PEPPER GRASS.)
 July—August. Flowers blue, yellow center.

DIOSCOREACEÆ. (YAM FAMILY.)

Dioscorea villosa, L. (WILD YAM.) July. Flowers whitish.

SMILACEÆ. (SMILAX FAMILY.)

Smilax rotundifolia, L. (BAMBOO.) June. Flowers small, greenish.
 tamnoides, L. May. Flowers greenish.
 Pseudo-China, L. (CHINA ROOT.) April—May.
 glauca, Walt. (SARSAPARILLA.) May.
 Walteri, Pursh. (RED-BERRIED BAMBOO.)
  March—April. Flowers brownish.
 lanceolata, L. August.
 laurifolia, L. July—August.
 auriculata, Walt. On the coast.
  May—June. Flowers small, very fragrant.

LILIACEÆ. (LILY FAMILY.)

Lilium Catesbæi, Walt. (SOUTHERN LILY.)
 August—September. Flowers scarlet, variegated with yellow and
 purple.
Yucca aloifolia, L. (SPANISH BAYONET.)
  May—June. Flowers white, tinged with purple.
 gloriosa, L. Sea coast. May—June. Flowers white.
 filamentosa, L. (BEAR-GRASS.)
  June. Flowers white, tinged with purple.
Allium VINEALE, L. (WILD ONION.) Introduced.
 striatum, L. March—April. Flowers white.

MELANTHACEÆ. (COLCHICUM FAMILY.)

Melanthium Virginicum, L. (BUNCH-FLOWER.)
 July—August. Flowers cream color, turning brownish.
Zygadenus glaberrimus, Michx. June—July. Flowers white.

Tofieldia glabra, Nutt. (FALSE ASPHODEL.)   October. Flowers white.
   pubens, Ait.   September. Flowers greenish white.
Pleea tenuifolia, Michx. Savannas near Cypress swamp, east from Wilmington.   October. Flowers greenish without.

JUNCACEÆ. (RUSH FAMILY.)

Juncus effusus, L. (BOG-RUSH.)   May—September. Flowers green.
   setaceus, Rostk.   May—July. Flowers green.
   maritimus, L. Brackish marshes. (*L. acutus, Muhl.*)
                                April—May. Flowers green.
   tenuis, Willd.   May—June. Flowers green.
   dichotomus, Ell.   May—June. Flowers green.
   polycephalus, Ell.   July—September. Flowers green.
   paradoxus, Meyer.   July—September. Flowers green.
   acuminatus, Michx.   July—September. Flowers green.
   marginatus, Rostk.   July—September. Flowers green.
Luzula campestris, D. C   March—April.
Cephaloxys flabellata, Desv. (*Juncus repens, Michx.*)
                                    July. Flowers greenish.

PONTEDERIACEÆ. (PICKEREL-WEED FAMILY.)

Pontederia cordata, L. (PICKEREL-WEED.)
                                July—September. Flowers blue.

COMMELYNACEÆ. (SPIDERWORT FAMILY.)

Commelyna communis, L. (DAY-FLOWER.)
                                July—September. Flowers blue.
   Virginica, L.   May—September. Flowers blue.
   erecta, L. Ger. McCarthy.   August—September. Flowers blue.
Tradescantia Virginica, L. (SPIDERWORT.) Eagle's Island causeway.
                            March—May. Flowers blue.
   rosea, Vent.   March—August. Flowers bright rose color.

MAYACACEÆ. (MAYACA FAMILY.)

Mayaca Michauxii, S. & E. Found at McIlhenny's mill, southern end of Wilmington, in bog by roadside. Causeway in abundance.
                            June—October. Flowers pale pink.

XYRIDACEÆ. (YELLOW-EYED-GRASS FAMILY.)

Xyris brevifolia, Michx. (YELLOW-EYED GRASS.)
April—May. Flowers yellow.
Caroliniana, Walt. July—August. Flowers yellow.
ambigua, Beyr. July—September. Bracts light brown.
tenuifolia, Chap. July—September. Flowers yellow.
torta, Smith. July—September. Flowers yellow.

ERIOCAULONACEÆ. (PIPEWORT FAMILY.)

Eriocaulon gnapholodes, Mich.
April—May. Flowers white, densely downy.
decangulare, L. (PIPEWORT.) July—September.
Pæpalanthus flavidulus, Kth. (YELLOW PIPEWORT.) April—May.
Lachnocaulon Michauxii, Kth. (HAIRY PIPEWORT.) May—June.

CYPERACEÆ. (SEDGE GRASSES.)

Cyperus compressus, L. July—September.
diandrus, Tor. Ger. McGarthy. August.
flavescens, L. July—August.
Gatesii, Torr.
rotundus, L. (NUT-GRASS.) [*C. hydra.*] August—September.
var. **hydra, Gray.*** Ger. McCarthy. August—September.
filiculmis, Vahl. (*C. mariscoides.*) July—September.
Michauxianus, Schultes. August—September.
Nuttallii, Torr. July—September.
flavicomus, Mich.* Ger. McCarthy. May—September.
Haspan, L. July—September.
speciosus, Vahl.
retrofractus, Torr. Ger. McCarthy. July—September.
stenolepis, Torr. August—September.
Baldwinii, Torr. Ger. McCarthy. July—September.
strigosus, L. July—September.
tetragonus, Ell. August—September.
vegetus, Willd. September.
vireus, Mich. July—September.
erythrorhizos, Muhl. July—September.
distaus, Pursh.* Identified by PROF. BRITTON as a long doubted species. Ger. McCarthy, 1885.

---

*Not given in Curtis' Catalogue.

Kyllingia pumila, Mich.   July—September.
Lipocarpha maculata, Torr.   July—September.
Dulichium spathaceum, Rich.   August—September.
Fuirena squarrosa, Mich. (UMBRELLA-GRASS.)   July—September.
Eleocharis equisetoides, Torrey.   July—September.
    cellulosa, Torr.* Ger. McCarthy.   August—September.
    quadrangulata. (*Scirpus quadrangulatus, Michx*) July-- September.
    tuberculosa, R. Br. (*S. tuberculosus, Michx.*) March—September.
    simplex, Torr. (*S: simplex, Ell.*)   May—September.
    prolifera, Torr.   May—September.
    rostellata, Torr.
    melanocarpa, Torr.   June—September.
    olivacea, Torr. Sea coast.   August—September.
    microcarpa, Torr.
    tricostata, Torr.   May—September.
    palustris, R. Br. (*S. palustris, l.,*)   June—September.
    obtusa, Schultes.   June—September.
    acicularis, R. Br.   June—September.
    pygmæa, Torr. Near the coast.   April—May.
    Baldwinii, Torr.   June—September.
Scirpus debilis, Pursh.
    Canbyi, Gray.* Ger. McCarthy.   July—August.
    pungens, Vahl. (SWORD-GRASS.) Near the coast. June—September.
    Olneyi, Gray. Brackish marshes.   June—September.
    lacustris, L.   July—September.
    maritimus, L. Salt marshes.   August—September.
    Eriophorum. Michx   July—September.
    lineatus, Michx.   June--August.
Eriophorum Virginicum, L. (COTTON-GRASS.)   June—August.
Fimbristyli-spadicea, Vahl. (*Scirpus castaneus, Michx.*)
                                                                 August—October.
Trichelostylis autumnalis. (*S. autumnalis, L.*)   July—October.
Isolepis ciliatifolia, Torr. (*S. ciliatifolius, Ell.*) Ger. McCarthy.
                                                                 August—September.
    stenophylla, Torr. (*S. stenophyllus, Ell.*) Ger. McCarthy.
                                                                 August—September.
    capillaris, R. & S. (*S. capillaris, L.*)   June—September.
Rhyncospora plumosa, Ell. (TICK-SEED GRASS.)   June—July.
    corniculata, Gray,* Ger. McCarthy.
    oligantha, Gray.   June—July.
    rariflora, Ell.   June—July.
    Torreyana, Gray.   July.
    microcarpa, Baldwin.   July—August.

---

*Not given in Curtis' Catalogue.

Inexpansa, Vahl.	July—August.
caduca, Ell.	August.
miliacea, Gray.	June—July.
Grayii, Kth	June—July.
megalocarpa, Gray.	May—August.
Baldwinii, Gray.	June—July.
ciliata, Vahl.	June—August.
fascicularis, Nutt.	June—July.
var. distans.	August—September.
filifolia, Gray.	July—August.
pallida, M. A. C.	June.
alba, Vahl.	August—September.
cephalantha, Gray.	July—August.
Chapmanii, M. A. C.	July—August.
Ceratoschœnus macrostachyus, Gray. (HORNED RUSH.)	August.
corniculatus, Nees.	July—September.
Psilocarya rhynchosporoides, Torr. (BALD RUSH.)	July.
Cladium effusum, Torr. (SAW-GRASS.) [*Schœnus effusus, Swartz.*]	
	July—August.
mariscoides, Torr. (TWIG-RUSH.) [*S. mariscoides.*]	
Dichromena latifolia, Bald.	May—July.
leucocephala, Michx.	August—September.
Scleria triglomerata, Michx. (NUT-RUSH.)	June—August.
laxa, Torr.	August—October.
Carex bromoides, Schk. (SEDGE GRASS.)	March—April.
stipata, Muhl.	April—May.
Muhlenbergii, Schk.	
stellulata, Good.	
straminea, Schk.	
fœnea, Muhl.	
crinita, Lam.	
polytrichoides, Muhl.	
granularis, Muhl.	
glaucescens, Ell.	July—August.
verrucosa, Ell.	
lupulina, Muhl.	July—August.
turgescens, Torr.	
Elliottii, Schw. & Torr.	
riparia, M. A. C.	
stricta, Good. (*C. acuta, Ell.*)	
laxiflora, Lam. (*C. anceps, Willd.*)	
debilis, Michx.	
vulpinoidea, Michx. *C. multiflora, Muhl.*)	July—August.

*retroflexa, Muhl.
tentaculata, Muhl.
†triceps, Michx.
*virescens.
folliculata, L.   (C. Xanthophysa, Wahl.)

GRAMINEÆ.   (GRASSES.)

Leersia oryzoides, Swartz.   (RICE GRASS.   FALSE GRASS.) July—August.
    Virginica, Willd.   July—August.
    hexandra,‡ Eagles' Island causeway. Ger. McCarthy. July—August.
Zizania aquatica, L.   (WILD RICE.)   July.
    miliacea, Michx.   (WILD OATS.)   April—May.
Hydrochloa Carolinensis, Beauv.   (*H. fluitans.*)   July—August.
Alopecurus GENICULATUS, L.   (FLOATING FOX-TAIL.) Introduced.   April.
Polypogon MARITIMUS, Willd.   (BEARD-GRASS.) Sea coast.
Sporobolus junceus, Kth.   (WIRE-GRASS.)   April—May, often October.
    Floridanus, Chap.§ Ger. McCarthy.   September.
    INDICUS, BR.   May—September.
Vilfa aspera, Beauv. (*Sporobolus asper, Kth., Agrostis clandestina, Spreng.*)
                                                      July—August.

Agrostis ALBA, L.   (BENT-GRASS.   HERD'S GRASS.)
        [*A. vulgaris, With., var. alba, Vasey.*] Introduced.
    elata, Trin.   (TALL THIN-GRASS.)   September.
    scabra, Willd.   (HAIR-GRASS.'   [*Trichodium laxifolium, Ell.*]
                                                   June—July.
    verticillata.§ G. McCarthy.
    perennans, Tuck.   (THIN-GRASS.)   [*T. perennans, Ell.*]
                                                July—August.
Cinna arundinacea, L.   (WOOD REED-GRASS.)
Muhlenbergia diffusa, Schreb.   (NIMBLE WILL.)   August—September.
    capillaris, Kth.   (HAIR-GRASS.) Near the coast.
Calamagrostis coarctata, Torr.   (REED.   BENT-GRASS.   WILD OATS.)
        [*Deyeuxia Nuttalliana, Vasey.   C. Nuttalliana, Steud.*]
                                           August—September.
    arenaria, Roth.   (*Ammophila arundinacea, Host.*) Sea beach.
                                                     August.
Stipa avenacea, L.   (FEATHER-GRASS.)   April.

---

*Is assigned in CATALOGUE to the " mountains."

†Is assigned in CATALOGUE to " Middle District."

‡Not given in Catalogue.   Possibly introduced from Florida or Gulf region.

§Not given in Curtis' Catalogue.

Aristida lanata, Poir. (Three-awned Grass.)   July—August.
    purpurascens, Poir.   August.
    stricta, Michx. (Wire-grass.)   June—July.
    gracilis, Ell.   August.
    virgata, Trinius.
    spiciformis, Ell.   August.
    oligantha, Michx.   September.
Spartina juncea, Willd. Sea coast.   July—August.
    cynosuroides, Willd.* Ger. McCarthy.   July—August.
    polysta.hya, Willd.   August—September.
    glabra, Muhl. (Marsh Grass.) Salt marsh.   August—September.
Gymnopogon racemosus, Beauv.   September—October.
Eustachys petraea, Desv. Sea coast.   May—August.
Cynodon dactylon, Pers. (Bermuda Grass. Reed Grass. Cane Grass.)
Ctenium americanum, Spreng. (Lemon Grass.) Savannas.
    July—August.

Dactyloctenium Ægyptiacum, Willd. (Egyptian Grass.)
    [*Eleusine* (?) *cruciata, Ell., E. Ægyptiaca, Pers.*]
Eleusine indica, Gært. (Goose-grass.)
Leptochloa polystachya, Kth. Brackish marshes.   September.
Triplasis Americana, Beauv. (Sand Grass.) [*Uralepis cornuta, Ell.*]
    August—September.
    purpurea, Chap. (*U. purpurea, Nutt.*)   August—October.
Melica mutica, Walt. (Melic Grass.)   April.
Glyceria nervata, Triu. (*Poa parviflora, Pursh.*)   July.
Arundinaria gigantea, Chap. (Cane.) [*A. macrosperma, Michx.*]
    February.
    tecta, Muhl. (Reed.)   February—March.
Brizopyrum spicatum, Hook. (Spike-grass.)
    (*Distichlis maritima, Raf. Uniola spicata, Ell.*)
    August—September.
Poa annua, L. (Spear-grass. May-grass.)   February—March.
    flexuosa, Muhl. (*P. autumnalis, Ell.*)   May.
    pratensis, L. (Blue-grass.)   May.
    compressa, L.   May.
Dactylis glomerata, L. (Orchard Grass.)   May—June.
Eragrostis Purshii, Schrad. (*Poa tenella.*)   June—September.
    tenuis, Gray. (*Poa tenuis, Ell.*)   August—September.
    pectinacea, Gray.   August—September
    var. refracta. (*Poa refracta, Ell.*)
Festuca Myurus, L. (Fescue-grass.)   March—April.
    tenella, Willd.   February—April.
    duriuscula, L. Sea coast.   April—May.

*Not given in Curtis' Catalogue.

ELATIOR, L.

nutans, Willd.　　　　　　　　　　　　　　　　　　August.

Bromus secalinus, L.　Ger. McCarthy.

RACEMOSUS, L.*　Ger. McCarthy.

Uniola paniculata, L.　(BEACH GRASS.)　　　July—August.

gracilis, Michx.　　　　　　　　　　　　　July—August.

Hordeum pusillum, Nutt.　W. M. Canby. (?)

Elymus Virginicus, L.　　　　　　　　　　　July—August.

Trisetum palustre, Torr.　　　　　　　　　　March—April.

Danthonia spicata, Beauv.　(WILD OAT-GRASS.)　　June—July.

Holcus LANATUS, L.　(VELVET GRASS.)　Introduced.

Anthoxanthum ODORATUM, L.　(SWEET-SCENTED GRASS.)　Introduced.
　　　　　　　　　　　　　　　　　　　　　　　April—May.

Phalaris intermedia, Bosc.　(SOUTHERN CANARY-GRASS.)　April—May.

Paspalum Walterianum, Schultes.　[P. vaginatum, Ell.]　July—August.

distichum, L.　　　　　　　　　　　　August—September.

præcox, Walt.　[P. lentiferum, Lam.]　　　May—June.

læve, Michx.　　　　　　　　　　　　　July—August.

Floridanum, Michx.　　　　　　　　　August—September.

undulatum, Poir.　[P. purpurascens, Ell.]　　September.

ciliatifolium, Michx.　[P. dasphyllum, Ell.]　June—September.

Panicum SANGUINALE, L.　(CRAB-GRASS.)　[Digitaria sanguinale, Scop.]
　　　Introduced.　　　　　　　　　　　　May—October.

gibbum, Ell.　[P. strictum.]　　　　　　July—September.

filiforme, L.　[P. striatum.　Digitaria filiformis, Muhl.]
　　　　　　　　　　　　　　　　　　August—September.

Curtisii, Chap.

agrostoides, Spreng.*　Ger. McCarthy.

hians, Ell.

colonum, L.*　Ger. McCarthy.

anceps, L.　　　　　　　　　　　　　August—September.

virgatum, L.　　　　　　　　　　　　August—September.

amarum, Ell.　On the coast.
　　　　　　　　September.　Plants salt and bitter to taste.

proliferum, Lam.　　　　　　　　　　　September.

divergens, Muhl.　　　　　　　　　　　August.

verrucosum, Muhl.　　　　　　　　　　September.

scoparium, L.　[nervosum.]　　　　　　　May.

viscidum, Ell.　　　　　　　　　　　　May.

scabriusculum, Ell.　　　　　　　　　　May.

depauperatum, Muhl.　　　　　　　　　June.

ignoratum, Kth.　　　　　　　　　　　July—August.

CRUS-GALLI, L.　Introduced.　　　　　August—September.
　　　var. hispidum.

---

*Not given in Curtis' Catalogue.

Setaria verticillata, **Beauv.** Introduced.
 glauca, Beauv. (Foxtail.) Introduced.
  var. lævigata. Ger. McCarthy.
 italica, Kth. (Italian millet.)   July—September.
Cenchrus tribuloides, L. (Sand-spur.) On the coast. July—October.
 myosuroides, H. B. K.* Ger. McCarthy. Ballast heap. West
 Indies.
Stenotaphrum* Americanum, Schrank.   June—September.
Tripsacum dactyloides, L. (Gama-grass.)
      August—September. June 20th, 1886.
Andropogon scoparius, Michx. (Broom-grass) August—September.
 furcatus, Muhl. (*A. provincialis, Lam.*)   September.
 tetrastachyus, Ell. (*A. dissitiflorus, Michx. var. tetrastachyus,*
  Hack.)   September.
 Virginicus, L. (*A. dissitiflorus, Michx.*)   September—October.
  var. vaginatus.
 macrourus. Michx.   September.
 Elliottii, Chap.   September—October.
Erianthus alopecuroides, Ell. (*E. saccharoides, Michx.*)
             September—October.

 contortus, Ell.* Ger. McCarthy.
Sorghum avenaceum, Chap. (Indian Grass.)
  [*Chrysopogon avenaceum, Benth., Andropogan avenaceus, Michx.,*
  A. ciliatus, Ell.]   September.
 saccharatum, Pers.* Ger. McCarthy.
 halapense, Pers. (Cuba Grass. Maiden Cane.)
 nutans, Gray. (Wood-grass.)
  [*Chrysopogon nutans, Benth., Andropogan nutans, L.*]
           September.

Lapago racemosa. Ger. McCarthy.

FLOWERLESS PLANTS.

FILICES. (FERNS.)

Polypodium incanum, Swartz. (Tree Polypodium.)
Pteris aquilina, L. (Brake.)
Adiantum, Capillus-Veneris, L. (Maiden's hair Fern.)
 Found by Mr. W. M. Canby, at Hilton, 1868.
Woodwardia angustifolia, Smith.
 Virginica, Willd. (*W. onocleoides, Willd.*)
Asplenium ebeneum, Ait.
Onoclea sensibilis, L. (Sensitive Fern.)

---

*Not given in Curtis' Catalogue.

Lygodium palmatum, Swartz. (CLIMBING FERN.)
Found in Holly Shelter swamp, 1879, by Maj. W. L. Young.
Osmunda regalis, L. (FLOWERING FERN.)
    cinnamonea, L.
Botrychium lunarioides, Swartz.

LYCOPODIACEÆ. (CLUBMOSS FAMILY.)

Lycopodium alopecuroides, L.
    Carolinianum, L.
Selaginella rupestris, Spreng.

HYDROPTERIDES. (WATER FERN FAMILY.)

Azolla Caroliniana, Willd.

NOTES OF THE BALLAST PLANTS.

BY GERALD McCARTHY.

Ranunculus philonotis is from southern Europe. Also found at Philadelphia by Mr. Martindale.

Schizandra coccinea is West Indian and Mexican, new to this country.

Scolymus Hispanicus is West Indian—new.

Tribulus terrestris is West Indian and also occurs in Southern Russia. Adventive at Philadelphia. Martindale.

Erodium cicutarium is European—new.

Linaria spuria is European—new

Heliotropium Europæum is European—new. Adventive in many places.

Datura mercifolium is West Indian and Mexican—new to this country.

Boerhaavia viscosa is West Indian and Mexican—new to this country.

Alternanthera achyrantha is West Indian and Mexican—new to this country. This plant is not a ballast plant but was found by me along the W. C. & A. R. R. in Columbus county, near Whiteville.

Euphorbia Cyparissias is European—new to this country.

Cyperus distans is an indigenous plant, but extremely interesting, having been found before by but one collector, Fred. K. Pursh.

Cyperus papyrus is the Egyptian paper reed, came probably from Sicily, where it grows abundantly.

Panicum colonum is European, and new to this country.

Cenchrus myosuroides is West Indian and Mexican, and new to this country, except the southern extremity of Florida.

Lapago racemosa is West Indian and Mexican, new to this country.

NUMERICAL STATEMENT OF GENERA, SPECIES AND VARIETIES, INCLUDED IN THE CATALOGUE.

PHÆNOGAMIA.*

Genera included in the Catalogue...............................	527
Species supposed to be indigenous to the Wilmington Flora.......1046	
Species regarded as introduced 122	
Total number of species...................................	1168
Varieties indigenous and introduced.............................	34
Total number of species and varieties	1202

These arranged systematically are distributed as follows :

Exogens ... .104 orders, 402 genera, 861 species and varieties.
Endogens .... 25 " 125 " 341 " "
Total Phænogamia ..129 " 527 " 1202 " "

COMPOSITION OF THE LARGER ORDERS.

FIGURES IN PARENTHESES REPRESENT INTRODUCED SPECIES.

ORDER.	GENERA.	SPECIES.		SPECIES AND VARIETIES.
Compositæ	56	(9)	142	146
Gramineæ................	49	(22)	118	121
Cyperaceæ	18		106	108
Leguminosæ	33	(12)	80	87
Scrophulariaceæ	16	(4)	31	34
Ericaceæ.................	15		29	31
Rosaceæ	11	(4)	28	28
Labiatæ	18	(8)	25	25
Onagraceæ	6		21	22
Umbelliferæ	12	(1)	22	22
Orchidaceæ	12		19	20
Caryophyllaceæ	13	(6)	19	19
Ranunculaceæ	6	(1)	16	17
Rubiaceæ	9		16	17
Cruciferæ	7	(3)	11	11

*This enumeration does not include the partial list of ballast plants given on page 134.

NUMBER OF SPECIES IN THE LARGER GENERA.

FIGURES IN PARENTHESES REPRESENT INTRODUCED SPECIES.

Carex	24	Ludwigia	11
Cyperus	20	Eupatorium	10
Rhyncospora	19	Hypericum	10
Panicum	(2) 19	Polygala	10
Solidago	16	Juncus	9
Aster	(1) 16	Coreopsis	8
Eleocharis	16	Lobelia	8
Polygonum	13	Utricularia	8
Desmodium	12	Asclepias	8
Euphorbia	(2) 12	Scirpus	8
Quercus	12	Smilax	8

Of varieties the catalogue gives two to Solidago, and one each to Cyperus, Rhyncospora, Euphorbia and Quercus of the above list.

CRYPTOGAMIA.

ORDER.	GENERA.	SPECIES.
Filices	9	11
Lycopodiaceæ	2	3
Hydropterides	1	1

Errata.

Page   3, line 26 from bottom, in place of " 1832" read " 1834."

"   11, line 5 from top, the generic name "Sarracenia" is accidentally repeated.

"   11, " Cakile maratima " should be " Cakile maritima."

"   16, the generic name " Hibiscus " is accidentally repeated.

"   18, line 5, the generic name " Rhus " is accidentally repeated.

"   18, line 6 from the bottom, " STAPHYLLACEÆ " should be " STAPHYLEACEÆ."

"   19, " Cassia Marylandica " should be " Cassia Marilandica."

"   20, " Lathyrus paluster" should be " Lathyrus palustris."

"   21, 'Styloshanthes " should be " Stylosanthes."

"   22, " Petalastemon " should be " Petalostemon."

"   26, " Sanicula Marylandica " should be " Sanicula Marilandica."

"   30, in note †, at bottom of page, " steets" should be "streets."

"   33, " Leucothœ " should be " Leucothoë."

"   36, "BROOM-ROPE FAMILY" should be "BROOM-RAPE FAM-ILY."

"   37, " Gratiola spœrocarpa " should be " Gratiola sphœrocarpa."

"   43, " Chenopodina maratima " should be " Chenopodina maritima."

"   34 — Note to S. Caesine refers to S. aspaloon.

Index to Genera.

www.ingramcontent.com/pod-product-compliance
Lightning Source LLC
Chambersburg PA
CBHW032347020726
47499CB00009B/3197